PATRICK & BEATR
MOUNT
MILWAUKEE

W9-ANJ-246

The Fight Over the Future

Science of Aging (SAGE) Crossroads (www.SAGECrossroads.net) is the premier online forum for emerging issues of human aging. At SAGECrossroads.net visitors are able to see, hear, and interact with prominent experts in aging and medical research, and with far-sighted thought leaders in bioethics, future studies, and health economics. From live debates on controversial topics such as cloning and health care rationing, to thought provoking articles on the latest science headlines, SAGECrossroads.net is a distinct, balanced forum where critical discussions and sharing of ideas can take place with the immediacy of the Internet.

Science of Aging (SAGE) Crossroads is funded through the generous support of:

The Archstone Foundation

The Archstone Foundation is a private grant-making organization whose mission is to contribute toward the preparation of society in meeting the needs of an aging population.

The Atlantic Philanthropies

The Atlantic Philanthropies seek to bring about lasting changes that will improve the lives of disadvantaged and vulnerable people.

The Retirement Research Foundation

The Retirement Research Foundation is the nation's largest private foundation devoted solely to serving the needs of older Americans and enhancing their quality of life.

SAGE Crossroads is an initiative of:

Alliance for Aging Research
2021 K Street NW
Suit 305
Washington DC 20006
www.agingresearch.org

American Academy for the Advancement of Science (AAAS)
1200 New York Ave. NW
Washington DC 20001
www.aaas.org

PATRICK & BEATRICE HAGGERTY LIBRARY
MOUNT MARY COLLEGE
MILWAUKEE, WISCONSIN 53222

The Fight Over the Future

✦

A Collection of SAGE Crossroads Debates That Examine the Implications of Aging-Related Research

Volume I, 2003

Sage Crossroads

iUniverse, Inc.
New York Lincoln Shanghai

The Fight Over the Future
A Collection of SAGE Crossroads Debates That Examine the Implications of Aging-Related Research

All Rights Reserved © 2004 by Alliance for Aging Research

No part of this book may be reproduced or transmitted in any form or by any means, graphic, electronic, or mechanical, including photocopying, recording, taping, or by any information storage retrieval system, without the written permission of the publisher.

iUniverse, Inc.

For information address:
iUniverse, Inc.
2021 Pine Lake Road, Suite 100
Lincoln, NE 68512
www.iuniverse.com

ISBN: 0-595-31631-X

Printed in the United States of America

305.26072
F471
2004

Contents

Introduction

Once considered a relative backwater amid the flowing information rivers of modern biology, the study of aging, or gerontology, has recently undergone a dramatic transformation. Thanks to a flurry of advances, most notably the discovery of simple genetic changes capable of dramatically extending life in model organisms, the field has begun to generate considerable buzz. It has even started scaring people, as repeated feats of life extension in animals have suggested the possibility of similar achievements in humans, thus prompting questions that earlier had been strictly the stuff of science fiction. *Will people ever live to be hundreds of years old? Is immortality in the cards? Will it be available only to the rich? How do we prevent hucksters from exploiting our fear of death to make a quick buck?* As the eight SAGE Crossroads debates and interviews published here demonstrate, today such questions are not at all out of place in the serious deliberations of scientists, bioethicists, and political theorists.

This introduction provides useful context for the exchanges that follow—beginning with background on the unique online forum that SAGE Crossroads provides. "SAGE," in this case, stands for "science of aging" (the connotations of wisdom and sagacity are, we hope, more than incidental). Two important Web sites now bear this name. The first is *Science* magazine's SAGE KE (http://sageke.sciencemag.org), a "knowledge environment" that covers research about aging and helps keep scientists in the field connected. The second is SAGE KE's younger sister site, SAGE Crossroads (http://www.sagecrossroads.com), freely accessible to all. Launched in early 2003 as a joint project by the publishers of *Science* magazine and the non-profit Alliance for Aging Research, Crossroads explores the public policy ramifications of the scientific discoveries chronicled on SAGE KE.

It does so in two ways. First, SAGE Crossroads publishes weekly news and expert observations about the social impact of increased understanding of aging. Second, it hosts monthly Webcast debates and interviews featuring leading scientists, ethicists, and journalists who have established a national reputation for thinking about the science of aging and its implications. These Webcasts provide a public airing of dialogues and disputes that had previously been confined to scientific journals, gerontology conferences, and related academic meetings. They have been widely accessed. On average, a lecture-sized audience of some sixty people tunes in for the live Webcasts. Afterward, thousands more access the debate videos and transcripts.

Thanks in part to SAGE Crossroads, then, it seems safe to say that more people than ever before have been spurred into thinking hard about where the science of aging may take us. But the Web site represents just one part of a broader intellectual convergence. The President's Council on Bioethics, chaired by Leon Kass, recently devoted a chapter of its report on the human implications of biotechnology, *Beyond Therapy*, to the ethics of age retardation. Meanwhile, political scientists such as Robert Binstock (featured in the sixth debate published in this book) have also called for a wider public discussion of current and foreseeable advances in the science of aging. Back in 1997, the world reacted with shock and alarm upon learning that a sheep had been cloned in Scotland. With enough advance public dialogue, perhaps we will be better prepared for, say, the achievement of dramatic age reversal–becoming physiologically younger–in laboratory mice.

The first two debates published here—one between *Reason* magazine science correspondent Ron Bailey and Johns Hopkins University political theorist Francis Fukuyama, and the other between UCLA Program on Medicine, Technology, and Society director Gregory Stock and celebrated environmental writer Bill McKibben—provide a sweeping introduction to the core ethical divide over life extension. In both debates, the differences between the two sides have been stated in fairly sharp terms. Ideally, these dialogues should be read alongside the Webcast interview with University of Michigan gerontologist Richard Miller, who lays out the basic facts about scientists' current ability to extend life in a range of organisms. Miller provides the scientific background; Bailey, Fukuyama, Stock, and McKibben then riff upon it with their ethical analyses. In the process, they provide a kind of sneak preview of a coming political brawl over human life extension that many scientists and ethicists think will take America by storm sometime in the next few decades. Reading through these debates today almost feels like eavesdropping on the future.

Representing the libertarian-leaning optimistic camp, Bailey and Stock regard the extension of human life as little more than the latest development in medicine's quest to cure disease. Therefore, they argue, we should welcome the increases in human happiness and improvements in quality of life that it promises. As Bailey put it in his debate with Fukuyama, "It takes more than a little hubris to believe that you are wise enough to tell other people to reconcile themselves that disease, disability, and death is the best thing for them."

Fukuyama and McKibben represent the more pessimistic side. Both fear the transformations that might result should human age retardation or reversal be achieved. As they see it, these include both unprecedented social disruption and a suspected loss of meaning itself as human life takes on a form and contour never before seen on Earth. Going all the way back to the epic of Gilgamesh, the pessimists have much of myth and literature on their side. "I take some comfort in my worries," observed McKibben in his debate with Stock, "when I consider that those people throughout history who have considered it most closely, who have done the deepest literary attempts to understand what it would mean to approach some sort of immortality and quasi-immortality, have come away scared, and beyond scared—horrified—as if they have peered into the abyss."

Despite their differences, however, both of these camps share a key assumption: human life extension *is* on the horizon. Yet with many issues in the biology of aging unresolved, it's crucial to note the existence of a third camp. Many scientific experts remain unconvinced that dramatic human life extension will be possible, due to the biological complexity of aging and ongoing uncertainty over whether a unitary aging process exists. Consider the debate between Richard Sprott, executive director of the Ellison Medical Foundation, and Aubrey de Grey, a gerontologist at the University of Cambridge. Even as de Grey predicts dramatic age reversal in mice within the space of ten years, and rapid advances in humans shortly thereafter, Sprott protests, "the human organism is enormously complex, and we don't know enough to override our genetic heritage."

A similar rift between scientific optimists and pessimists emerges in yet another debate published in this book—between Harvard Medical School pathologist Roderick Bronson and Jackson Lab senior staff scientist David Harrison. The debate centers on whether so-called "biomarkers of aging"—physiological traits that, when measured, would reflect the biological rate of aging in a living organism—exist. Bronson considers the quest to discover biomarkers foolhardy, arguing, like Sprott, that there's simply no unitary aging process to measure. Harrison, on the other hand, considers the starkly different aging rates across mammalian species evidence that aging might be subject to relatively simple genetic controls, and that the right biomarkers could reflect this process in action.

Where a scientist stands on the question of biomarkers appears to correlate closely with whether that scientist thinks we'll manage to modify the human life span anytime soon–and thus, whether the ethical concerns raised by McKibben and Fukuyama should trouble us. We hope that the publication of this book, along with continuing coverage and debate on SAGE Crossroads, will help us to identify the proper balance between speculative worries about the future on the one hand, and a realistic assessment of what that future may hold on the other.

The remaining debates and interview published here concern more immediate issues. The ethics of cloning human embryos for biomedical research purposes, or so-called "therapeutic cloning," lies at the center of a contentious, no-holds-barred debate between Advanced Cell Technology CEO Michael West (who's pro-cloning) and *Washington Post* columnist, and medical doctor, Charles Krau-thammer (who's against it). West then becomes a subject of discussion himself during the Webcast interview with Stephen Hall, author of the recent book *Merchants of Immortality*, which covers stem cell research, cloning, and other controversies. In a later debate, University of Illinois at Chicago demographer S. Jay Olshansky and Case Western Reserve University ethicist Robert Binstock discuss the present day antiaging movement and its promotion of unproven and potentially dangerous herbal and hormonal treatments.

Stem cell research and therapeutic cloning, often grouped together under the heading of "regenerative medicine," should not be confused with age retardation or age reversal. If successful, these regenerative technologies could allow us to extend human life indirectly by growing replacement organs and battling degenerative diseases. Such advances would not, however, impact the *rate* of human aging itself—which means that in some sense, they present a far more logical extension of current medical science, rather than a radical departure. Nevertheless, embryonic stem cell research and therapeutic cloning have received far more public attention thus far than age retardation and age reversal have. Scientific issues aside, this situation has a great deal to do with the fact that both involve the destruction of human embryos, and have therefore become caught up in America's interminable battles over the politics of abortion.

But any and all of these technologies, if successful, would face the potential for rampant commercialization. As Stephen Hall notes in his interview and in his book, "immortality" sells. So strong is the human desire to avoid aging or cheat death, in fact, that members of the popular antiaging medicine movement *currently* claim the ability to slow human aging through the use of various untested dietary supplements.

In response, a group of fifty distinguished gerontologists have gone on the offensive, publishing statements in *Scientific American* and elsewhere debunking prominent antiaging claims—the subject of the last debate in this book. At this point it remains unclear whether the *Scientific American* campaign will achieve its goal, or whether it will instead leave the public confused and unable to distinguish between mainstream gerontologists—many of whom believe that significant life extension will someday happen by medical means—and their "antiaging" opponents. S. Jay Olshansky defends the position that the war on antiaging medicine has been successful, but Robert Binstock argues for a different, more conciliatory approach. Gerontologists, he says, "are getting inextricably involved with a tar baby."

These debates and interviews share a deep moral seriousness and public mindedness. Granted, readers will notice some bristling at times between ideologically disparate intellectual combatants. But rather than fearing such clashes, we should welcome dialogue on these topics all the more for its rawness and intensity. The future won't wait. Better to fight about it now than later.

What are the Possibilities and the Pitfalls in Aging Research in the Future?

Francis Fukuyama, SAIS
Ron Bailey, *Reason* Magazine
Morton Kondracke, Moderator
February 12, 2003

Pictured: Dan Perry, Executive Director of the Alliance for Aging
Research; Morton Kondracke, moderator; Francis Fukuyama, SAIS; and
Ron Bailey, *Reason* Magazine.

For more information on debate participants and SAGE Crossroads go to
www.sagecrossroads.net

KONDRACKE: Today's debate is formally entitled "The Future of Aging: Pitfalls and Possibilities," but I think an informal subtitle would be "Brave New World or Fountain of Youth?" because our two discussants have very different views about where aging research is leading us—and they are distinguished commentators indeed.

Francis Fukuyama is the dean of the faculty and Bernard Schwartz professor of international political economy at the Paul H. Nitze School of Advanced International Studies at Johns Hopkins. He has written widely on issues concerning democratization in the international political economy, and he is, I guess, most famous of all for his book *The End of History and the Last Man*, published in 1992, which has been published in over twenty foreign editions. He's got a provocative new book out called *Our Posthuman Future: Consequences of the Biotechnology Revolution*, which raises the specter that we're headed toward that brave new world.

Also joining us is Ron Bailey, who is the science correspondent for *Reason*, the monthly magazine on politics and culture. He has written articles for *Reason*, including "Forever Young: The New Scientific Search for Immortality," and he's working on a new book to be called *Liberation Biology*, which we will get into. Previously he's done several weekly and documentary programs for public television and ABC News, and he has written for innumerable publications including the *New York Times Book Review*, *Smithsonian*, *National Review*, *Forbes* and *Reader's Digest*. Ron, welcome.

FUKUYAMA: Thanks very much Mort. I really appreciate being invited to speak, although I sense that I am being set up in a certain way as having to take the losing side of this debate. If this were the Oxford Union the question would be something like, Early death, disease, debility: pro or con? and I'd be asked to take the pro side. I think that there are a lot of things that can be said about biotechnology in general, but I want to focus particularly on the aging side since we are being sponsored by the Alliance for Aging Research, and I think that that raises some of the most difficult questions.

Now, as I said, I don't think I can seriously win this debate if it's posed in that form, and I'm not going to really try. I think everybody would like to live longer—they would certainly like that longer life to be healthier; I don't think anyone can possibly contest that. Although I've argued in my book and other places that we need to update and modernize our regulatory system to take account of some of the new technological developments in biotech in the area of aging, I don't think that is appropriate. So nothing I say should be construed as making a case for blocking anything or stopping or regulation, because I think that is simply not doable. All I want to do is say that the quest for life extension and the prolongation of life through biomedicine is simply not as unalloyed a good thing as I think some people think it is. We ought to think carefully about where we are headed. We may not be able to head off this particular future, but I think it's worth thinking about it.

Now there are, I guess, three general categories of concern. One is a very simple utilitarian one, which is that much of the value of life extension obviously depends on the quality of life that is being extended. I would say that although this is something that people don't like to address, all of the great advances in biomedicine up to this point that have brought us life expectancies up in the mid-eighties for women for example are themselves not an unalloyed good.

At the age of eighty-five, something like fifty percent of people develop some form of Alzheimer's, and the reason you have this explosion of this particular disease is simply that all of the other cumulative efforts of biomedicine have allowed people to live long enough to where they can get this debilitating disease. Now maybe we will cure it through further biotechnology, but the rate of progress in solving these age-related problems tends to be uneven. It would be nice if we could assume that in the future all of these technological developments will allow every system to keep going until all of a sudden—boom—one day we simply die, having lived as twenty-year-olds up until then. But I think that the probability of that this will happen is probably low. We can imagine some scenarios that are pretty grim, actually, where people would like to die and aren't able to.

I had a personal experience with this; my mother was in a nursing home for the last couple of years of her life and if you see people caught in that situation it's really a fairly morally troubling thing because nobody wants their loved ones to die, but these people are simply caught in a situation where they have lost control. So that is one thing—we don't know whether this is going to happen. It may be that future progress will be quite even and the scenarios won't materialize but it's something worth thinking about.

The second argument—and this should appeal to libertarians that take individual choice seriously—is really a question of the social consequences of life extension. Life extension seems to me a perfect example of something that is a negative externality, meaning that it is individually rational and desirable for any given individual, but it has costs for society that can be negative. I think if you want to understand why this is so, you just think about why evolution makes us, why we die in the first place, why in the process of evolution populations are killed off. I think it clearly has an adaptive significance, and in human society generational succession has an extremely important role. There is the saying among economists that the science of economics proceeds one funeral at a time, and in a certain sense a lot of adaptations to new situations—politically, socially, environmentally—really depend on one generation succeeding another.

For better or worse, people come into the world with a certain world view that is formed when they are born and they generally tend not to give that up at later stages of their life—even through a lot of political progress, a lot of scientific progress, a lot of adaptation to new circumstances. For example, the only people that vote Communist in the former Soviet Union right now are people that are over the age of sixty-five. I think there are a lot of issues like that that have to do with life cycles; things that may be desirable from an individual standpoint may have consequences, and certainly if that life extension is at a much lower level of activity or health, the economic consequences could be quite serious.

I guess the final set of issues has to do with the whole quest for immortality and what that suggests as a desirable human good in and of itself—apart from all of this consideration of social consequences. We have this kind of assumption that if a life of 78 years is a good thing, then a life of 140-160 years must be twice as good. It seems to me that is not obviously the case. If what drives this quest of immortality is a fear of dying, then in a certain sense it's a fool's errand because you are going to die sooner or later; you're simply putting off that moment and not facing that aspect of human existence.

Further, the real question is—and this is the area where I think I can least persuade anybody, but I myself take this quite seriously—what we live for and what we want those years for. "What we want those years for" in terms of quality, and by quality I don't just mean, are we healthy? It's a question of higher goals that we live for, not simply extending the life span.

I think you must consider that the noblest virtues in human societies have always been associated with people that have a certain relationship to death. For example, when taking into consideration people who are willing to risk their lives on behalf of causes greater than themselves, one has to ask the question, Well, what does this desire—the possibility of indefinite life extension—mean for that particular set of virtues and the whole possibility of facing death?

Mort mentioned my book was called *The End of History,* and it does seem to me that Nietzsche's "last man" would be subscribing to *Life Extension* magazine—too eager to continue his or her mediocre life for as long as possible without worrying about some of these higher questions about what that life is used for. But again, as I said, this is not an argument that I expect; it's not an argument really in favor of any particular policy. It really is simply meant to raise a different set of moral questions about what is really our objective in seeking these kinds of goals.

KONDRACKE: Thank you. I would not be as defensive as you think you have to be here; I think that you raise some very challenging points about what life is all about. We'll even get into these various social implications and political implications that you've raised—questions about whether science is in the purpose of changing human nature and our political structures as well—so don't feel as though you can't convince anybody.

Ron Bailey, you wrote in an article in *Reason* magazine that the defining political conflict of the twenty-first century will be the battle over life and death. On the one side stand the partisans of mortality—which I take it you would regard as the party of death, or if you want to be even more critical, you'd say the party that counsels that humanity should quietly accept our morbid fate and go quietly into that good night. On the other side is the party of life—where I gather you put yourself—that rages against the dying of the light and yearns to extend the enjoyment of a healthy life to as many people as possible for as long as possible. What is your thesis here?

BAILEY: Well, I'm definitely going to say that on the question of resolving who is in favor of death, debility and disease, I am on the con side of that. But the Alliance for Aging Research and AAAS, I thank you for having this discussion. The media advisory I got had the question, Are scientists playing God? and I note that on Sunday the *Washington Post* informed us what the answer was. They put God on notice.

In any case, I would like to deal with some of the same issues. My speech would be entitled "First, Do No Harm." I would like listeners to remember three things from my discussion. First, the point of aging research is not to enable us to be older longer, but rather to allow us to be younger longer. The goal is not to have a world filled with nursing home residents—and there may be transitional problems, which Francis Fukuyama has identified—but the goal is not to get older longer.

The second thing is that I would like people to remember that human beings are constituted by evolutionary history in such a way as to identify the potential problem first and to think about negative issues. The way a zoologist buddy of mine explained it when I was complaining: he said, "Ron, think about it this way, when we were growing up, evolving, in Africa, some guy would say 'there is a tiger in that tree is going to eat you.' You would pay attention to that and run away. But if he said, 'You know, there are some fruit trees over that mountain there and some benefits you could get,' you'd say, 'No, I'll think of that tomorrow.'" So we are constituted mentally to think about problems, and it is very hard to think about benefits. I think we can get into some of that later.

And then, as I said, my talk is "First, Do No Harm." There are definitely people, of which I don't think Francis Fukuyama is one, who do believe that biomedical research should not aim at lengthening human life spans. For example, Leon Kass, the president's favorite bioethicist, asserts, "The finitude of human life is a blessing for every individual, whether he knows it or not." Or there is Daniel Callahan, who is from a different political perspective—he's a co-founder of The Hastings Center. He has declared, "There is no known social good coming from the conquest of death." Yet also, of course, the worst possible way of resolving the issue is to leave it up to individual choice. So I ask, what life-lengthening research would such opponents ban? After all, treatments to prevent or cure Alzheimer's will likely have the side effect of normal brains functioning better longer. Treatments to cure or prevent heart or other circulatory diseases will also lengthen life. Treatments to prevent or cure cancer will certainly extend the span of our days. Can such people as Callahan or Kass seriously want to limit research for cures for Alzheimer's, heart disease and cancer?

But perhaps I am mischaracterizing the opponents. Perhaps what they really want to do is in some sense limit treatments that are aimed specifically at the process of aging itself—such as treatments that would reduce the damage caused by free radicals. But undoubtedly it will turn out to be the case that such therapies that reduce the damage caused by free radicals will be very useful in treating specific diseases such as Alzheimer's, heart disease and cancer. The fact is that biomedical research is almost by definition aimed at lengthening healthy human life, so again what research opponents think should be prohibited on the grounds that it illegitimately aims at lengthening healthy human life spans? I think they need to answer that question.

Let's take a very brief look at some of the social consequences of longer average healthy life spans. The first concern often mentioned is that longer life spans will contribute to the overpopulation problem. Keep in mind that the reason the world's population quadrupled in the twentieth century was not because people began breeding like rabbits, but because they stopped dying like flies. Demographers believe that the normal life expectancy of 1900 was around thirty years; today global life expectancy has more than doubled to sixty-six years. I ask: who can doubt that this increase in average life expectancy has resulted in the greatest mass improvement in human well-being and happiness in all of history?

Well what about the future? Won't much longer life spans cause massive overpopulation? University of Chicago demographer Jay Olshansky has calculated that if everyone on the planet were immortal tomorrow—we woke up and had the happy news that we're now immortal—while maintaining the current projected trends in current fertility, world population would rise to around thirteen billion in 2100. That number, he notes, is the same number the alarmists like Paul Ehrlich used to predict that would be for the middle of this century, by 2050. I suspect that the vast majority of people would be very willing to take on the problems of longer life spans and figure out ways to deal with them. I believe that one hundred years would give human society enough time to adjust to longer, healthier lives.

Another argument sometimes heard—we heard it here by Frank Fukuyama—is that the elderly have an obligation to future generations to die and get out of the way. Please remember that our ancestors did not ask us if it was all right for them to more than double their life expectancies; they just went ahead and did it because they knew there was a high human good to achieving that.

With regard to this notion that longer life spans are a negative externality for society, I should point out that at least Hobbes would argue that society was at least created for the purpose of enabling people to live lives that were not nasty, brutish and short; in other words, society is for the purpose of helping us live better and longer lives.

So what about getting out of the way of the younger generation, which is also a concern we also have heard raised here? Actually, our society is already organized to encourage just that. And I'm not talking about things like mandatory retirement. Professor Fukuyama worries about how to dislodge set-in-their-ways geezers clinging to the top positions in our society. But consider that Bill Gates didn't work his way up the ladder it IBM; he started his own company before the age of forty. Or, consider that biologist Craig Venter didn't wait to become head of the National Institutes of Health; he went out and created the Institute for Genomic Research, Human Genome Sciences, and Celera. In other words, our competitive society already makes sure that dead wood gets cleared out pretty regularly. And it's the same thing with our politics; we have regular elections where forty-somethings like Bill Clinton can become president.

Of course, there are likely social benefits in a world where people live longer. Already we see some. Families the world over are having fewer children because the ones that they have will likely make it to maturity. The average woman in 1960 had around 6 children over her lifetime, and locally that number is down to 2.7 and continuing to fall. Also, it's a truism that human beings are very short-sighted. However, if we can regularly plan on living more than a century, we may develop more foresight and wisdom about the long-range effects of our activities.

Opponents of biomedical research talk a lot about hubris. But it takes more than a little hubris to believe that you are wise enough to tell other people to reconcile to themselves that disease, disability, and death is the best thing for them. With regards to those opponents who assert that certain biotechnological research violates human dignity, I think that they owe us a more precise account of just what constitutes of violation of human dignity if no one's rights are violated. Dignity is a fuzzy concept and appeals to dignity are often used to substitute for empirical evidence that is lacking or sound arguments that cannot be mustered. After all, what is so dignified about dying of Alzheimer's, diabetes or cancer?

The president's bioethicist, Leon Kass, told the *Washington Post* earlier this week, "The pursuit of perfect bodies and further life extension will deflect us from realizing more fully the aspirations to which our lives naturally point, from living well rather than merely staying alive." I want to suggest to you that this is a false dichotomy. To live well one must first stay alive. So to answer the question posed here—Are scientists playing God?—I believe the answer is no. Human dignity is not a quality that depends on limiting human lives to a certain span of years. The highest expression of our human nature is to try to overcome the limitations imposed on us by our genes, our evolution, and our environment. The highest expression of our human nature is our quest to maximize individual human flourishing, alleviate physical and mental disease and disability, and lengthen healthy life spans. Future generations will look back at the beginning of the twenty-first century with astonishment that some very well meaning and intelligent people actually wanted to stop biomedical progress just to protect their cramped and limited vision of human nature. They will look back, I predict, and thank us for making their world a world of longer, healthier lives and for making that world possible for them. Thank you.

KONDRACKE: Well thank you very much for launching that, both of you. This is deeply provocative stuff. I have a feeling, though, that Frank Fukuyama would not say that we shouldn't develop cures for Alzheimer's disease; I think there is a larger philosophical question. We can get back to, Is the world going to be overcrowded? in a second, or to, How can we afford to take care of all these old people who are in retirement? and the social consequences.

Let me just raise the fundamental question of Frank's book, in which he says that in *Brave New World* Aldous Huxley argues that "we should continue to feel pain, or be depressed, or lonely or suffer from debilitating disease, all because that is what human beings have done for most of their existence as a species." Frank says that the aim of his book is to prove that Huxley was right; that the most significant threat posed by contemporary biotechnology is the possibility that we will alter human nature and thereby move us into a post-human stage of history. Human nature shapes and constrains the possible kinds of political regimes that we have, so a technology powerful enough to shape what we are will have possible malign consequences for liberal democracy and the nature of politics itself—which is the highest philosophical plane of discussion. Now what is this dystopia that you fear this biotechnology will produce?

FUKUYAMA: Well actually, there are several things wrapped up in that; the political concern that was alluded to in the passage you quoted really has to do with the possibilities for social control that these technologies offer. I think that one of the reasons we ended up with liberal democracy is that a lot of the ambitious social engineerings that were tried by various utopian regimes all failed, and they failed really because of human nature. In Communist society you wanted to abolish private property and family, and people just weren't engineered that way, so they resisted it. That's why we've ended up with a fairly benign outcome at the end of the twentieth century where liberal democracy is really the only game in town. But that's really dependent on technology, and if you have technology that in a much more scientific way understands the basic roots of behavior—and a lot of this comes out of cognitive neuroscience and is manipulable, not so much by genetic information but by things like neuropharmacology—that gives you a fairly powerful set of tools for one group of human beings to alter and control the behavior of others. I think in some sense you see this beginning, with the way we medicate young children with drugs like Ritalin.

The deeper issue is this whole question, which I admit to a lot of people seems fuzzy, of human nature and human dignity and the rights that come out of that. I would simply say that you can say it doesn't have a sound empirical basis, but virtually every one of us believes in it in some form or another. Our Constitution was founded on a respect for human beings as having certain inalienable rights that come from a certain set of essential characteristics that defines a human being. All the big political struggles in our history have been focused on who is admitted into this charm circle of people who have human dignity. African-Americans were excluded and women were excluded, and they are now allowed in because they possess those essential characteristics as much as white males do. But there is this human essence that in my view is really determined by the extremely complex nature that we have been given by the evolutionary process, and it's an essence that we don't understand the complexity or wholeness of.

The hubris is—I don't like the phrase "playing God"—the hubris is that we simply do not understand the complexity of our own social structures, and the attempt to engineer and to improve incrementally this thing and that thing for limited aims is almost inevitably going to have unintended consequences, some of which could be extremely serious. The idea that we can simply use technology to master this is hubristic. It is true in a lot of other areas. People thought we could damn rivers and produce electricity and didn't understand that there were huge environmental consequences as well.

KONDRACKE: Well what is your ultimate nightmare? *Brave New World* raises the possibility of "soma" where everybody is zoned out and therefore can't respond to any unpleasantness—and in fact have ruled out unpleasantness. I would think that one of the things that you would be afraid of would be genetic engineering, somehow tinkering with germ line cells to the point that you could manufacture different castes of human beings to be worker bees or rulers or something like that. Is that what you're talking about? We're talking *Blade Runner* kind of stuff. Do you think that is a realistic possibility, and is that what you are worried about?

FUKUYAMA: I think that kind of germ line engineering is much further off in the future—maybe Ron will have a different opinion—I think that is not going to happen any time soon for a number of reasons. But I think that is something important to worry about because if that technology does come down the road, you have at least one scenario that is troubling. That's actually greater uniformity because you get social trends that people follow where they don't want their kids growing up gay or different or something like that.

KONDRACKE: But you think genetic selection is a bad thing?

FUKUYAMA: It's not a bad thing per se and it's not the method that's questionable, it's really the uses to which it is put. Genetic selection and that whole range of technology assumes that we know what makes a better human being, and this gay gene business is a good example of that. If you could engineer that out out of a child's genome or prevent that chemical or biological predisposition to gayness would people choose this? Is it a power that they ought to have? And by doing that, they say to themselves, "Yes, I will improve my child as a human being." These issues are fraught with—there is a big normative dimension and its not so simple to say, "Yes we're going to make the human race better and they will be grateful to us later on for having done that."

KONDRACKE: Ron?

BAILEY: Sure, a couple things. One is that I know you were saying that we have a document which guarantees inalienable rights and among those inalienable rights are life, liberty, and the pursuit of happiness, which I don't think contradicts in any way the notion that we should have life extension or allow people to go ahead and...

FUKUYAMA: Well, who gets it? That is the important point. What creatures are entitled to have those rights? Because we don't give them, for example, to chimpanzees even thought they share ninety percent...

BAILEY: The fact of the matter is, we do make gradations in society already, as you well know, based on the capacity and there will be a standard where if you have a particular known capacity, you get the full set of rights. That is what we've been doing. That is what happens. Our human nature up until two hundred years ago warranted things like slavery, warranted things like women being chattel. Our human nature is not what made us better, it was a political decision over time because we learned more about what we were like and we continue to do that. I think that in fact what the technology does is enable greater levels for human beings. I think that's what we have to focus on with regard to genetic engineering.

FUKUYAMA: That still seems to just get around the whole question of what is this essence that we are trying to protect that is the foundation of our life. Every one of us implicitly believes that there is this creature that is endowed with these rights and science itself may give us more information as to the genetic information of different species and so forth, but it doesn't give us any guidance on that normative question of why would we afford superior rights to certain classes of beings, and that's the central issue.

BAILEY: Again, that's a political decision that we've made over time and we will continue to make those decisions. I agree with you; liberal democracy is probably the final state of human society, and I think that we are not going to move backwards. We may in fact expand the scope of rights, but we are not going to remove the rights.

KONDRACKE: Ron, if it were possible to define a gay gene would you be in favor of equipping parents with the right not to have gay children? Is it laissez-faire?

BAILEY: No. For that particular issue the way I come at the problem is a little bit different. What I am trying to do in some of my writing is find the kinds of things that human beings would want to have—capacities that they definitely would want to have. Let's face it, people would prefer to be more intelligent than they are; they would like to have a healthier immune system; they might like to avoid certain kinds of diseases. These are general capacities that other people naturally get—people already have this. I see no reason to deny parents access to technologies that can provide those capacities for their children as well. Now with regard to things like eye color and so forth, we do not say, "You are allowed to do that," because at least at this point it is too fraught with danger and too likely to be a fad. It may be well be that the gay gene problem falls on that side of the line. I haven't thought that through yet.

But again, I want to point out that we don't have to think through everything now because as our technologies incrementally advance, we learn that human beings are terrible at foresight. The problem is we always see dangers and we try to stop them. We ignore the benefits; we do what human society has done to advance: we try something new. We develop something and if it doesn't work out then we say, "Gosh, that didn't work out," and we move backward and there is no perfection to it. There is no way to in advance warn against all problems. My favorite example of technology with regard to foresight is the guy who invented the laser in the 1960s. He said, "I have no idea what this could possibly be used for," and look at it now. There are lasers everywhere now. We use them to fix our eyes; we use them to do surgery; we use them to run our printers. The fact of the matter is that technologies have lots of beneficial consequences that cannot be foreseen. It is a lot easier for us to imagine dangers than it is to imagine benefits, and I would like to stress that I think that most people will see benefits stemming from biotechnology.

KONDRACKE: Now one of Frank's theses in the book is that the government needs to step in and regulate biotechnology, and you seem to think, Ron, that there is a place for government to regulate technology. I have a feeling that you each would regulate different things or give the government certain powers that the other wouldn't. But what is it, Frank, that you would have the government stop? Is it therapeutic cloning? Is it genetic designing? Do you want to ban Prozac?

FUKUYAMA: Let me make a couple of points. First of all, we already heavily regulate biomedicine and we in fact ban Prozac for certain uses. We ban Ritalin for certain uses. We make a distinction between therapeutic uses of these drugs and biomedicine generally, and what you might call non-therapeutic or enhancement uses. So it isn't as if this is something new. And we've got this institution here in Washington called the Food and Drug Administration that already sets lots of limits. We could advance the speed of medical research vastly if we allowed doctors to run clinical trials, like the Nazis did where people were deliberately infected with agents, but we take a slower rate of advance because we don't like that kind of free experimentation for ethical reasons. So the precedent is that we regulate, and the question is, do we need to modernize that system of regulation? For example, the FDA has rather casually asserted an authority to regulate human cloning. It is not as clear that the statute allows them to do that or whether that assertion of authority would stand up under court challenge.

I'll just give you a very concrete policy example where regulation is necessary. If you want to have the cloning, I think what you need is a system like the one that now exists in Britain under the Human Fertilization Embryology Act. That act created a Human Fertilization and Embryology Authority that regulates embryo research in Britain. Every embryo that is produced has to be registered with the government, and they track what happens to it. There is a requirement that it actually be destroyed after fourteen days so you can't clone an embryo and then let it grow into a fetus and then harvest the organs or something of that sort. In fact, it facilitates research cloning in Britain because it provides firebreaks against the kinds of slippery slopes that unregulated pursuit of this kind of research would lead to.

KONDRACKE: How do you feel about that, Ron? Theoretically, to take the example of cloning, if you could have a fetus forming and use developing embryos, and if you could take stem cells, why not harvest hearts? What is your view of that?

BAILEY: Well, I'm against that. I should point out that the FDA, when we're talking about regulation, regulates for three things typically: safety, quality, and efficacy. We don't impose values on people. We don't say Prozac is bad for people to use. We say it should be safe, it should be high quality, and it should work, and that's why we have it and the fact is, those standards should remain. The troubling problem I have is creating an agency that would be able to impose uniform values on the rest of society. The values are what are a struggle for me, sitting around deciding for everybody else how they should live their lives and what a perfect version of human nature should be.

KONDRACKE: Let's go back to the aging issues and the issue of longevity. Frank, what Ron says is the purpose is not to simply extend life but to make life more healthy, and theoretically it would be possible to not have people living the last fifty years of their lives with Alzheimer's. Couldn't we figure out ways and things to do? There is lot of work to be done—taking care of children, helping the needy—and theoretically could we find useful work for people if they were robust long past one hundred in a democracy?

FUKUYAMA: Well sure. I think Ron is probably right that we will eventually adjust to this sort of thing. I think that a lot of it really does depend on exactly how robust those lives are and I agree that this aim of this research is not to be old for longer but the question is can we get it and no once can predict that, no one can control that process. Callahan and Kass are not saying, "Stop research on aging." They have particular positions on stem cells for other reasons, not because they want to stop work on aging, and I'm not saying that either. I'm just saying there is a kind of presumption in a lot of these debates that anything that prolongs life or cures a disease is automatically and necessarily a good thing and that that goal trumps any other kind of moral consideration, and I am just saying that is not self-evidently true.

BAILEY: First of all, Callahan does say, and I can actually cite—he points out in his last book that "no new medical technologies should be developed at all until all technologies are currently available and deployed to everybody." He also says nothing more than palliative treatment should be allowed to anyone over age seventy. So he's definitely against this, period. But with regard to things like technologies having problems, for example one, of the last guys who received one of these artificial hearts just died last week. The fact is that there will be people who will take these risks, and they should be allowed to take them. I think they know what they are doing, and their lives may not be the quality you would want, but the fact is that they are participating in the process of trying to make lives better for us all. Technologies are not perfect out of the gate, and again there will be a messy birthing process to all of these technologies, but because there will be problems is not a reason to stop them, and I think you agree with that.

KONDRACKE: There is a microphone over there. I invite you to throw question at the discussants. Please participate. We do have a couple of questions from the Web site…One is related to how the media covers science: "It's often difficult to discern hope versus hype. How do you balance the hope versus the hype in your beliefs regarding the progress of science in aging research?" In other words, are we being overly optimistic, are all these hopes that we are citing legitimate, or is there hype going on here?

BAILEY: Well, as a journalist, I've probably done my share of the hyping, but I think that actually these technologies are all possible and will likely come to fruition probably much less rapidly than we would like. I've just been doing some research by the way. One of the arguments you often hear is, "We have to have a societal discussion about all of these technologies." Well, I want to point out that with regard to cloning we've been having these discussions forever, starting out with *Brave New World*. But there was also an article on the front page of the *New York Times Magazine* in 1972 and they thought we'd be cloning people by 1980. The fact is that none of these technologies will likely be coming around as rapidly as I would like, but they will all be available by the end of the century.

KONDRACKE: Studio audience, yes, go ahead.

AUDIENCE MEMBER: First of all, I would suggest that the issue of immortality is a red herring. One thing I would say with regard to unintended consequences of research is that the one intervention in aging that we know that works is caloric restriction, and yet caloric restriction applied in mammals—which is the best model we have for humans—in fact really has no more effect than extending life by thirty or forty percent. The thing we do know about this is that in fact these animals are in better health. So they are not in poorer health longer, they are in fact in better health longer. They eventually die of the same things that rats or mice die of who are not treated with caloric restriction. So I really think it would be nice if we could get this issue of immortality out of this debate on aging research, because I really think it's not going to happen. If you extrapolate mouse and rodent results to humans what you're talking about is average life spans of 110 to 115, which we already know is the maximum that's been observed. I don't have a particular question, but I would be interested in any comments you may have.

KONDRACKE: What is the possible life span here that we are talking about? Immortality is probably off the charts for the foreseeable future, so what is it we are talking about and what are the consequences of a 110-year life span?

FUKUYAMA: Well, while I'm the one that used the word "immortality," I doubt that it's ever going to be possible. What I think is more important is what is driving the desire for longer life spans ultimately is something like a desire for immortality. Do people somehow think that it is more humanely appropriate that people live to be 150 rather than 70? Is there something magical about that number? No. Why not 300 rather than 150 or 600 rather than 300? The logic of this really is a desire to put off death and that is the only reason I use that word. I think it is a fool's quest in many respects to even think of that as a goal.

BAILEY: The way I prefer to think of the issue is we should try to make death optional. The problem is, the quest for immortality is not something that is new to human nature. The fact is that evolution threw up a creature that could reflect on the fact that it is going to die, and if we look at our cultures, I would maintain that our cultures are devised around two things: reproduction and death. One of the oldest myths we have comes down to us from ancient Sumeria and is the Gilgamesh saga where basically the hero is trying to find a way not to die. And what do you think pyramids and cathedrals are? Aspirations toward life, a different kind of life, because they couldn't imagine physical immortality; it was all too obvious that people died fairly frequently in the bad old days. And now I think

Frank Fukuyama is right; we are the desire here. The impulse here is the natural impulse is to live a long, healthy life, and perhaps forever, if possible.

KONDRACKE: Other questions. Yes, go ahead.

AUDIENCE MEMBER: A little bit before Gilgamesh is Cicero's essay on aging where Cicero is having a conversation with two younger citizens. They said, "Well don't you mourn the fact that you are not as vigorous and healthy as we are?" And he said, "But I have other opportunities at my stage in life," and I think that is something that we can't lose a perspective on. What value does an older citizen have and what can they contribute? Do we need to make them just younger and more like the twenty- or thirty- or forty-something? Is there something to sustain the value that they add at that stage in life? So your perspectives are complimentary, I believe, not so much a dichotomy of views. Do you have any comments on that?

FUKUYAMA: Well I do think that there is a certain logic to the human life cycle; that there are certain things that are appropriate when you're a child, and others when you are a young adult and starting a family, and that other virtues occur in later periods of life. One of the things that I find disturbing about this quest for the extension of life is that it completely interrupts that life cycle—a hypothetical case where you've got people not dying at all. Reproduction has to basically go away as a central issue in people's lives and I think that is profoundly unnatural. You may say, well what is so great about nature? but I do think that we all have a very strong attachment to these natural forms of living by which we believe we flourish as human beings.

KONDRACKE: Ron?

BAILEY: I actually don't see any contradiction at all. You're right. The fact is that we don't value people because they are physically old; we value them because of their experiences and the way they've shaped their lives and the advice they can give us. I don't think that will go away. In fact, I think that will get even better, that they will be able to provide us with greater wisdom over time if they stay vigorous and healthy into old age. I don't think that's a contradiction at all.

KONDRACKE: We are going to lose our Web connection right now. Thank you, those of you who have been on the SAGE Web site, for joining us. We'll have one more question afterward, but I think this has been a wonderful beginning to what is going to be a very rich series of discussions. We will be getting into more practical subjects in future discussions, as in, "Can we afford to have millions and millions of healthy old people in years to come?" But for those of you on the Web we are ready for sign-off, so thank you very much.

End.

Do We Want Science to Reinvent Human Aging?

Gregory Stock, UCLA
Bill McKibben, Journalist
Morton Kondracke, Moderator
March 27, 2003

Pictured: Morton Kondracke, moderator; Gregory Stock, UCLA; Bill McKibben, journalist.

For more information on debate participants and SAGE Crossroads go to
www.sagecrossroads.net

KONDRACKE: The formal title of today's debate is "Do We Want Science to Reinvent Human Aging?" but I might say that the informal title of the debate might be, "Do We Want Science to Reinvent Human Beings?" The person who says yes is Gregory Stock, and his book is entitled *Redesigning Humans: Our Inevitable Genetic Future.* Greg Stock is the director of the Program on Medicine, Technology, and Society at UCLA's School of Medicine, and in this role he explores critical technologies poised to have large impact on humanity's future and the shape of medical science. His goal has been to bring about broad public debate on these technologies and their implications on public policies.

Bill McKibben graduated from Harvard University in 1982 and became an editor at the *New Yorker* magazine; his work has appeared not only there but also in the *Atlantic,* the *New York Review of Books,* the *New York Times, Natural History, Rolling Stone, Esquire,* and *Audubon,* among many others. After five years at the *New Yorker* he decided to leave the frantic pace of urban life and move to an isolated house along with his wife in the Adirondack Mountains in New York State. Now he also spends time in Vermont, where he says he has cross-country skied for 116 days this season. Our question today is: do we want science to reinvent human beings? and Bill McKibben's answer is no. His book-length exploration of this subject, which was recently published, is called *Enough: Staying Human in an Engineered Age.*

Our format is that this debate is going to last an hour. Greg Stock will begin with a ten or twelve-minute exposition of his point of view, followed by Bill at the same length, then there will be some back and forth which I'll try to encourage with questions. We'll go on until we get so many good questions from here in the studio audience or out there in cyberspace that we can just do the rest of the program answering your questions—so we invite your questions. Those of you who are watching this online, there is a place on the Web site, which should be clear to you, on how to e-mail your questions, and we'll be glad to have you participate. With that, let Greg Stock begin.

STOCK: Thank you. It's good to be here. Bill McKibben and I have very different visions of the future. We don't differ so much about the significance of the possibilities emerging from the revolution of biotechnology today, as about how to navigate the terrain ahead. I think that we have to be extremely cautious about regulating these technologies, and that legislation based on abstract fears about the future would be a big mistake. Moreover, I believe that we already have adequate structures in place to protect against the immediate dangers at hand, and that we already have adequate mechanisms in place to respond to the kinds of challenges that are likely to arise from the various concrete problems that will undoubtedly emerge as we use these technologies.

We have to consider the extraordinary nature of what is going on today. We are unraveling the fundamental workings of biology, coming to understand what makes life tick. Of course we're going to try to use this knowledge in ways that we think will enhance our lives. Of course there are going to be applications of this knowledge that will be troubling and challenging to us and that will create a lot of angst. These are profound developments. They're going to revolutionize health care and medicine; they're going to change the way we have children; they're going to alter the way we manage our emotions; and they are probably going to alter the human life span. These are very, very challenging developments, and I think that the possibility of altering the aging process itself provides an excellent example of the larger challenges ahead. We could be talking about designer children, or we could be talking about psychopharmacology, but I think aging is the most interesting, because if we in fact are able to retard, and perhaps even reverse, some aspects of this process, the impacts upon us will be so enormous.

I don't think any other technological advance will affect our lives as profoundly, and I think it's very likely that this will occur. Indeed, I'd like to nudge it forward a bit. If I have a choice of being among the first generation to experience the benefits of extended human health span and prolonged vitality, I'd prefer that as opposed to being among one of the last generations not to have the advantages of those possibilities. And I suspect that Bill McKibben would come down on the opposite side of that.

Before looking at aging in particular, though, let's step back and take a look at where we are. It's important to do that, because otherwise it would seem that there really might be a possibility that we could just say "enough" and stop these technologies. But what is happening is unprecedented in the history of life, and there is no brake to put on. There are two revolutions occurring; the first is the silicon revolution. We've seen the Internet and telecommunications, and the way these various technologies are changing our lives. We have taken the inert sand at our feet, the silicon, and we have breathed a level of complexity into it that rivals that of life itself. This is a breakthrough that is going to mark a transition in the history of life, and it will always be seen as such. The second revolution is a child of that development, and it will have even more impact on us. This is the revolution in biological sciences, the genomics revolution, the unraveling of life. Essentially we are beginning to intervene in that realm and take control of our evolutionary future, which is obviously going to have huge impacts. Bill's idea of now saying "enough" is an interesting hope, since we're really just beginning.

Let's return to aging and consider the consequences of interventions that would retard or even reverse key aspects of it. The questions are easy and lots and lots of people have asked them and worried about them: Is this going to divide society? Is it going to bring about the loss of our values? Is it going to create intergenerational conflict? Lead to a population explosion? We can throw away all of these questions, though, because the answers are unknowable at the present. They depend on technologies that still are in the future and will be brought to bear on these issues. Aging is multifaceted, and different aspects of it are going to be dealt with in different ways in different cases. So would treatments for aging have side effects that are serious? Will they need frequent repetition? Are they going to be arduous? Are they going to take a long time to act? Are they going to have to begin before a certain age, perhaps even in embryo? Are they going to reverse aging or simply slow it? You can capture a number of these possibilities by looking at the fixed and variable costs, the general development and overhead costs of the interventions as opposed to the costs per person for the procedures. If antiaging interventions emerge, are they going to be the kinds of things that can be rolled out to large numbers of people relatively inexpensively? If so, then there will be huge political pressures to make this happen. Or are they going to be very high-cost and very personalized and very difficult? in which case there will be a much narrower subset of the population that has access to them, which would present a host of other problems.

Not everything that can be done should or will be done. But when you take something like this that is likely to be feasible in thousands of laboratories throughout the world, something that is seen as beneficial by large numbers of people—which is clearly the case with antiaging medicine, judging from the way people do cosmetic surgery and vitamins and such—something that's almost impossible to police (and in fact how are you going to punish somebody? Are you going to force them to eat French fries and smoke cigarettes?), it's not a question of *if*, but *when*, and where and how this will arrive.

I think it's absolutely certain that humanity is going to go down this path for two reasons. The first is that it's just the natural spin-off of all the mainstream medical technology that virtually everyone supports, and secondly because we're human. We try to use technology in order to enhance our lives in a variety of ways, and to imagine that we won't go down this path if we can is every bit as much a denial of what the past tells us about who we are as it would be to think we would go down this path without agonizing about it and fretting a great deal. I think that there will be a lot of that as well, and that Bill is a good example of this. The lines are blurring between therapy and enhancement, between treatment and prevention, and between need and desire, and they are going to continue to become ever more blurred as we move forward.

We could try bans; we probably will in a variety of realms, but ultimately this is not going to stop this development, it's merely going to drive it from view, preserve the technology for the wealthy, shift it overseas, and even more importantly, raise the dangers from these technologies by denying us the information we need to use them wisely. There are going to be mistakes, there are going to be challenges, and it would be best to make those when very few people are involved, rather than later when they are being rolled out broadly.

It's critical to realize that you can't just sit back and reflect on these sorts of possibilities and think that you're going to come up with the best path forward. You have to buy this knowledge, and you have to buy it through experience, through mistakes, through dealing with the various challenges that arise as we begin to use these new technologies. The critical thing is to put the mechanism in place that will enable us to obtain this information as rapidly as possible. These new technologies are no ticket for utopia, but I think that the benefits far outweigh the risks, that humanity will explore these realms, and that our next frontier is not space, but our own selves. To me the challenge facing us is not how we handle genetically modified foods or cloning or any other specific technology, like anti-aging medicine, it's whether we will continue to have the courage to embrace the possibilities of the future or whether we pull back out of fear and essentially turn over their development to other braver peoples in other regions of the world—many of whom, of course, will have values that are very different from our own.

It's important to realize that there will be no consensus about these issues. Lots of people say, "Well, let's just slow down and debate and discuss it until we reach consensus." It's not going to happen. These things touch us too deeply. They touch our values too deeply. They depend upon religion, upon philosophy, upon politics, upon culture, upon personality. In fact, some people are going to continue to see them as the invasion of the inhuman—the worst possible thing that could happen to humanity—and other people are going to say "this is the flowering of human endeavor, this is incredible, this is stuff that previous generations could only dream of." Of course we're going to go out and use this.

It is very important in the debate to realize that this is not like nuclear weapons, where someone is going to misuse the technology and billions of people are going to be injured, or even vaporized. As long as we guard our political institutions and freedoms, we will have a way of protecting ourselves from biotechnology.

The biggest danger is not that we're going to move ahead too rapidly, but that because of misguided legislation, we will actually injure large numbers of people indirectly by delaying might-have-beens that will come too late for those today with serious diseases. It's very easy to dismiss these injuries-of-omission associated with misguided policies, because they are a bit in the future and it is not quite clear who's going to be affected, but it will be very clear to those people at that time, and we should not ignore this.

KONDRACKE: Thank you very much Greg. Bill McKibben, you have time for your response, and I hope that in your response—I know that this is a key part of your book—you will explain what some of these technologies are that Greg alluded to, and perhaps at least describe them for the benefit of those in the audience in lay language.

MCKIBBEN: Absolutely. Thank you Mort, and thank you Greg. It's a pleasure to be here, although, I must say to be on the side of death and aging seems a hard task. I think that Greg Stock has written a fascinating book on these questions and done much interesting speaking and thinking about them and is being too modest in his description of this whole idea that we should just wait and see what happens and what comes down the road next. I mean, it's very clear what people are talking about and what he's talking about. He's made it clear in his work that the intents to reverse and conquer aging will be the first things that may well tempt us to do germ line engineering of human embryos. Other people working on these issues have said much the same thing and talked about cloning of human body parts—like Michael West, the CEO of Advanced Cell Technology—the kind of immortality and things like that.

As I say, this all feeds into one of the oldest dreams of human beings: the idea that we might, if not completely conquer our mortality, certainly delay it dramatically. In fact, many of the people doing this work are all for conquering it entirely. I take some comfort in my worries about this when I consider that those people throughout history who have considered it most closely, who have done the deepest literary attempts to understand what it would mean to approach some sort of immortality and quasi-immortality, have come away scared, and beyond scared—horrified—as if they have peered into the abyss.

Let it be said that we are not talking about cancer treatments, etc., that kind of thing. I mean, if we are, then we have no grave debate between us. Those things that would treat diseases to which we now fall victim and that would attempt to make us healthier through what is now the normal course of human life hold no problems for me. The interesting question is whether or not we should attempt to extend human life beyond the sort of life span that we see now, a maximum life span of something like 115 years—what scientists and biologists have called the Hayflick limit, the number of cell divisions that happen before we in essence wind down. It is that that we really and deeply would consider and that professor Stock considers in great depth in his work.

First of all let me say just in passing—we can get back to this—that the idea that this will be the first thing to tempt us into using germ line engineering is an important thing not to pass over quickly. One of the points in my book is that all of the uses of germ line engineering lead us into potentially deeply troubling acts. In fact, by their very nature they lead us automatically into troubling acts. Professor Stock has proposed that people might want to redesign their children so that they will be more religious or more musical or more optimistic or things like that. I think down that path lies the death of what we call human meaning, the idea that people are in some way their own human beings and are not pre-programmed semi-robots of some kind.

But let's talk more directly about aging, because abolishing death, or at least dramatically postponing death, is a very good example. I'm willing to concede to Professor Stock and his allies that it may well be possible at some point. But if it does become possible, it will be the most fundamental shift imaginable in the course of human history. It will not be a small extension of some other thing; it's not like developing a treatment for Alzheimer's or some other condition, it's something of an entirely different order. We call ourselves mortals for good reason; in many ways that is our single most important defining characteristic—more important than the opposable thumb or the idea that we are the creature that knows that someday we will perish upon this earth. Without it, it would be a world where we live farther and far longer, where we, in Professor Stock's vision, double, triple, quadruple, maybe sextuple our life spans or where we're able to clone new organs and keep ourselves alive forever—to use the more far-out visions of some of the nanotechnologists and the people really interested in what Professor Stock has called the "silicon revolution." If the day comes when we can ensure some kind of immortality through some link with that silicon universe, then we will be something completely and utterly different, and with that shift will come, I think, a kind of meaninglessness. In a life like that, time would have very little meaning. The idea that for each thing that there is a season, or that there are choices to be made, would begin to disappear; the rhythms that have marked our lives would begin to disappear, and the sadness is not those things themselves, but that with them they will take much of the possibility for human maturation, which much more than anything else is the proper role of the human life.

The arresting of aging is in some ways the arresting of maturation; the destruction of that force, our own knowledge of our transience, that leads us in some ways to grow up, to not put ourselves at the center of the universe forever. In its wake it leaves, I think, a kind of pervasive selfishness. Let me quote briefly from Dr. Michael West, the CEO of Advanced Cell Technology, who is on the cutting edge of this sort of work. Asked about these sorts of questions of immortality, he outlined for his interviewer a long list of the many things that would make it possible, and in his view desirable, for this to happen. The interviewer asked him about whether or not it would create practical problems in terms of things like population. He alluded as to how it would; in fact, the world would begin to fill up and all of you can figure out the practical problems, I don't want to dwell on them. But a universe in which you spend eighty percent of your life belonging to the AARP, would be a crowded universe eventually. I don't want to focus on the issue because I'm not sure what the answer is, but what interested me was the way that he said it: "Why put the burden on people now living, people enjoying the process of breathing, people loving and being loved?" The answer is clearly to limit new entrants to the human race, not to promote the death of those enjoying the gift of life today. Now, today. He and his colleagues want to stop time, but you can't enjoy the gift of life forever. Maybe with these tools we will in some way learn to live forever, but the joy of it, the meaning of it, will melt away like ice cream on an August afternoon. I guess what I am trying to say is that I think that life far beyond the parameters of what we know now, life that goes beyond the normal human expectations, may be very much like a trap, and the name of that trap is a very American one—the constant idea that more is better. If it is good to live 80 years, it must be better to live 180 years and far better yet to live 300 years.

In the first place, just from a completely utilitarian point of view, it's not entirely clear that this even leads to more. In many ways more is not always better. There are questions that are thresholds instead. I mean, we're going to go out to dinner tonight and if we have a beer or two that would be pleasant, and if we have eight or nine that would be a mistake. There are plenty of cases in which more is not better, and one of the facts of growing up is beginning to realize which things fall into this category. I'll end it there so that we get to questions, except to say that I think that one thing not worth debating, not an important part of this debate at least, is the idea that things are inevitable. If things are inevitable, then there is no reason to debate them or write books about them, have forums, have Web casts, interrupt the course of our afternoon, whatever else. It's the very fact that things aren't inevitable that makes them interesting; the very fact that we need to debate them. Professor Stock said at one point that the consensus will never emerge; if by consensus we mean an absolute unanimous agreement about things, that's obviously true. But, if as in a democracy, we mean a working majority who will pass legislation and act responsibly, then clearly over time in some way or other that could emerge, and that's precisely the direction to which eventually all this should be heading.

KONDRACKE: Professor Stock, would you explain to our audience, for the benefit of those who don't understand it, briefly and as simply as you can what germ line genetic engineering is and any other emerging technologies that you think are relevant to this issue? I know that you refer to something called "fyborgs," which are like bionic cyborgs but they're different, that's an area of exploration that you talk about. So what are the technologies exactly and how imminent are they?

STOCK: Before I respond to a few of those points, the kind of technologies that we're talking about in a larger sense are things called germ line engineering, which is the alteration of the germinal cells—as in the germination of a seed—so it's actually going into the first cell of an embryo and altering the genetics of that cell. And that's not so distant, though it is probably at least a generation or so away. Another analogous technology, and it exists today, and it again has to do with embryos, is to screen them. You take a six-cell embryo, remove one of the cells, run a genetic test of that cell, and depending on the result, decide whether to implant the embryo or discard it, and that's here today, at least for a number of genetic diseases. We'll soon be able to make screenings based on personality traits and non-disease traits. There are all sorts of pharmaceutical interventions that one could imagine to alter aspects of aging as we understand the biological basis of it, but the specifics of future technologies is something we can only see with very blurred vision. We can't see very far, and so that's why there is a bit of non-discreteness here about what we're talking about.

I want to respond to a few of the points that Bill made. I think they show one of the problems associated with this important discussion. And the discussion is important, because it is a way of getting us to grapple with these technologies, and not to decide whether we can stop them or not. If we determine that these developments are going to happen—and I think that is very clear—there are still are many different paths open to us, and some of them will be much more painful to us than others. So it matters if we think that we can turn back these technologies by blocking embryonic stem cell research or other new technologies.

Bill may say, "Well I'm not talking about any sort of disease, I agree with those sorts of things, even though I disagree with extending our life spans," but these two realms cannot be separated because as we unravel the processes of life in biomedical technology—which is the way we are going to fight diseases—we open up all sorts of other possibilities. Some critics might assert that the only way to block these possibilities is to try to sift through the research and find the developments likely to have substantive impacts on biomedicine today. This is the situation with embryonic cell research.

The discussion of immortality is very interesting because of how quickly people move from the idea of expanding life span or health span to suddenly gaining immortality with all of the attendant trauma. You can make an argument, if you're going to talk about human immortality, that if you triple human life span you haven't even gotten close because 85 over infinity is about the same as 250 over infinity. Face it: immortality is something far beyond what we're talking about.

KONDRACKE: What do you think the practical reasonable limits are? How close to immortality is it thinkable to get?

STOCK: If you did not suffer any aging whatsoever and had the vitality of a teenager, then deaths from accidents alone would bring life spans of about one thousand years. So, visions of immortality are not a good foundation for real-world public policy discussions.

KONDRACKE: A thousand is pretty close.

STOCK: That is if you had no aging whatsoever. If you're going to say, well, we don't want to extend the human life span because of problems of population growth and various other issues that might arise, or problems of the human spirit, then I think you really have to examine whether in fact you should be working on extending the human life span at all by treatments of Alzheimer's and all sort of diseases of aging. There is a blurring between treatment and prevention, and between therapy and enhancement. If you in fact could alter the aging process so that people live a bit longer and do so in a healthier fashion, well that is going to look very much like preventative medicine for the diseases of aging. But bit-by-bit, we are likely to move towards greatly extended life spans. The two are going to be essentially equivalent, so the question is not whether we're for immortality, but whether we're comfortable with healthier, longer lives.

MCKIBBEN: Leonard Hayflick, who you know is the godfather of aging research, said not long ago that if you manage to cure all the causes of disease that are on every death certificate there was, you would extend human life expectancy less than we have in the course of the last century through public health measures.

STOCK: Less than forty years.

MCKIBBEN: It would be much more on the order of twenty years. If you got rid of cancer and heart disease, then you've only expanded average human life expectancy five to six or maybe seven years. Those sort of things I don't think people have—I don't know anyone who has great difficulty with them. People, say, at conferences you've organized, have talked about living to be a thousand, or indefinitely, and about crossing over to some new world, and to pretend that that is not something one should take seriously or think about is to muffle the impact of your own work.

STOCK: First of all, Leonard Hayflick's idea that the maximum number of divisions that a cell can undergo has much to do with organismic aging in higher animals is not something that most people that I know in biology of aging subscribe to. So this distinction is kind of a red herring.

MCKIBBEN: What most people agree is that 115 or so is where we max out.

STOCK: There is actually a lot of disagreement about that. There is certainly an aging process that is occurring. But what I'm trying to say is that it's great to talk about immortality, but in fact we're dealing with real lives in the real world. So when you say, "Well, we don't want to be immortal," what does it mean? If you're just saying "Well, I have a lot of angst about that and I think we should relinquish these technologies and not think about them or not support them publicly or whatever," that's one thing. I don't think it will have much of an impact on medicine, but it's a great discussion to have. But if you're saying, "Well, I really want to stop these technologies because I'm afraid of immortality," the only way you can really do that is to intervene in the realm of biomedical research that is directed at real people with real diseases today.

MCKIBBEN: Let's talk about things like germ line engineering, which most countries in the developed world have now decided they don't want to pursue and have had some bans.

STOCK: And you would ban it?

MCKIBBEN: I would, but I wouldn't ban stem cell research. I think that stem cell research is being held hostage at the moment by people who want to go ahead and eventually do germ line engineering and who are unwilling to work out the compromises necessary to let it go forward in a good way. That is to say, it's completely possible to imagine stem cell work where you look closely, regulate it closely enough that human reproductive cloning is not much of a danger. But working out that kind of a system in some way, like, say, what the President's Council on Bioethics has proposed doing, is precisely what a lot of researchers have refused so far to do.

KONDRACKE: I thought, though, from reading your book that you are against therapeutic cloning?

MCKIBBEN: No, you thought wrong. What I'm against is going down that path before we put in place all the safeguards necessary to make sure that it doesn't become reproductive cloning.

STOCK: Well how could you put in all the safeguards that are necessary? The technological process is basically the same.

MCKIBBEN: That's right, you'd need to have very, very careful regulation and monitoring of all the work that is being done, just in the same way that we've been having a chemical industry without also producing chemical weapons.

STOCK: So let me be more concrete. You've said you don't like the idea of germ line intervention, which is operating on an embryo. Now we've been talking about aging—and we can get back to germ line interventions because I see that as being a natural outgrowth of a whole number of other screening procedures. In my view, if you're going to repair genes or select an embryo that does not have a defective gene, it's an engineering choice. But let's set that aside and talk about antiaging medicine. It's possible that germ line interventions will be necessary to greatly extend the human life span. But that's not really very interesting to most people who are already adults because they've passed the stage where you can intervene in the embryo. So most of the research of the Michael West's of the world are directed toward adults, quite simply because he and others would like to see benefits that could be applied to themselves, and most people are beyond the embryonic stage. So what would you propose as things they could actually do to stop this, other than the rash of legislation surrounding stem cell research?

MCKIBBEN: Let me say first of all that your notion that germ line engineering is somehow a small part of this or some instance—the second sentence of your chapter on aging says, "In light of our yearnings for immortality the underlying biology of aging may well be the first germ line intervention to truly tempt us," because everything else will have only very, very modest effect.

KONDRACKE: Which is to say that you would discover what the genetic causes of aging are, and you would tinker with the genes for ever after in embryos and eliminate the aging process for that baby and all that baby's progeny.

STOCK: Putting genes in place that could be turned on and off at a later time by the adult to operate against some of the processes of aging might be the only possible way of intervening.

MCKIBBEN: You predicted it would.

STOCK: I think that such an approach will be used, but you're jumping over a whole mountain of effort that has to do with antiaging medicine in adults. If it turns out that you need to do embryonic interventions in order to gain significant life extension, that will be profoundly disappointing to virtually all of the people that you have cited, and to those doing the kind of work to which you are referring. I'm not against germ line intervention, I think that it will occur—but what I'm saying is that as for antiaging medicine, the work that you believe might lead to germ line intervention, how would you stop it?

MCKIBBEN: I'm not here to write the series of regulations. There are people who are more adept at that. This series had Frank Fukuyama a few weeks ago; this is his specialty. I am interested in the sort of deeper question of whether or not this is a path that we want to go down. If all we're going to talk about is whether or not we're gong to do some more work on Alzheimer's, then I don't think there's going to be much debate. Why don't you want to engage these questions that you've opened up so dramatically and powerfully in your own work? Questions about this future where we're engineering people not to age, we are engineering people so that they will be more religious, or we're engineering people so that they'll be more optimistic, or we're engineering people so that they'll be more musical or faster or smarter?

KONDRACKE: In addition to the aging issue you would say, I think, that it is fine that we develop basically super-people where everybody's kids would be Michael Jordan.

STOCK: I think that that is a misunderstanding of what enhancement is likely to be. In that people always think of super-humans and blond, blue-eyed sort of rendition of the Nazi eugenics?

KONDRACKE: I was thinking Michael Jordan.

STOCK: OK, in Europe they might talk about this, these evocations of Hitler. But what I think is most likely to occur is that yes, parents will make choices about personality, about temperament. Let's take real simple ones that are available today: Parents can make choices, and do make choices, about the sex of their offspring. Now I don't know what you're feeling is about this, but Francis Fukuyama and many, many other people say that it is inappropriate, that a parent should not be able to make those sorts of choices. But I say, if a couple wants to have a boy or a girl, for some particular reason, who is a child of the gender they choose, and equally, if a couple feels that they would have a certain resonance with a child of a particular predisposition—let's say they have a tendency to sleep through the night—I don't see a problem with that, although you can argue that this would somehow corrupt the relationship between parent and child. I would say, well, what about birth control? I mean, we have profoundly altered the family using birth control. We have essentially separated reproduction and sex, and nobody talks about that. I think it can be seen in very much the same way.

MCKIBBEN: I think that argument, that deciding when you are going to have a child is the same as deciding whether or not that child can be programmed to be pious or not, is a logical leap that would defy super-human.

KONDRACKE: Bill, let me ask a question here. Do you oppose genetic screening?

MCKIBBEN: I don't—with a set of caveats. When confined to a set of genetic illnesses that society decides is worth screening for, then I don't oppose it.

STOCK: You would oppose selection for gender.

MCKIBBEN: I would. And I would oppose selection for intelligence and blondness and height and all those things as well. I think down that path lies a very different society in the ways that people have talked about divisions in society between rich and poor; that's all true, and all things that Francis Fukuyama has written about are very true and telling. But at an even deeper level, it presupposes an entirely different idea of what a human being is and what the relationship between human beings is going to be.

One of the things in my experience being a parent, one of the glorious things about it and one of the things about it that causes you to become a fuller human being, to mature, is that at some level you weren't in control of this process, that the person who emerged was not your catalog choice of things. You didn't go into a clinic and say, "I want…" The notion that this is a good idea and that people will go for it is perhaps at least in a few cases true. I mean we live in a society where everybody is busy injecting botulism toxins to wipe out wrinkles. On the other hand, if we think about it a little, we'll see the kind of logical fallacies. For instance, if you engineer your child today, and in fact Professor Stock sort of alluded to this, you give them the best gene package available, up their IQ twenty points and then research goes on. Ten years later you're ready to have your second child and there's a whole other set of better software available to plug in. Now you're giving them forty IQ points and their muscle mass is twice as big. Well, your first child is now just like Windows 95. What kind of world do you enter into when you begin to do this?

STOCK: It's very interesting because you assume that you're going to really be able to do dramatic enhancements beyond the envelope of what is considered human potential today. I think that that may come, but that is a technologically challenging development that's very difficult.

MCKIBBEN: You were the one who explained how Bill Gates' child in thirty years would be able to engineer far better than anything the richest person on earth could afford now.

STOCK: I said that divides are likely to occur between one generation and the next, just like with technologies of computers or such, and that the richest person in the world could not purchase anything that was even akin to what anyone can get today.

KONDRACKE: But there is no question, is there, that rich people would have this technology available to them? The richest people would ensure that their kids went to Harvard by raising their IQ in vitro and maybe get an athletic scholarship to boot, and poor people would be left behind and would be some sort of subspecies.

STOCK: No, the distortion here is basic. I made that statement about Gates, but I did not say that he would be able to choose superhuman traits. That is the hardest thing to accomplish. Let's just take something very simple to illustrate this. It's very difficult to imagine, but if you were going to try and engineer a human being that was eight feet tall, it would be very difficult because you'd have all sorts of physiological problems, things that wouldn't work well. But if you wanted to increase the height of someone, say a boy who was going to be five feet tall or four foot six, and bring him up toward average height or a little bit above average, that would be relatively easy. So these kinds of interventions in the foreseeable future are much more likely to be applied to those at the lower end of the spectrum of performance in any particular realm just because it's so much easier technologically.

MCKIBBEN: That represents a willful misreading of the society in which we live. If you look around the world, the idea that medicine and medical technology is applied where it's most needed is one of the grave misunderstandings of our time. If that were the case, we'd be spending billions of dollars fighting malaria and not baldness and erection problems.

STOCK: Take erection problems if you want to. I'm saying it is deficiencies that people will direct their energies toward because the incentives are so much higher. That's true with virtually every intervention that you can imagine. It is generally the sick and the ill and those who feel they are diminished in some way or another who are willing to accept the risk and burdens of these kinds of interventions, it's not someone who has an IQ way above average who is going to subject himself to an uncertain intervention to try to gain a few more points.

MCKIBBEN: Most of human growth hormone is being sold to parents of children who in fact were not going to be dwarves but were going to be just a little shorter than average.

STOCK: Shorter than average. It's how they would perceive it. In their terms. When you see these interventions, they do slip into the other realm toward enhancement, of course they do.

KONDRAKE: If you are seriously talking about some day being able to produce people who are a thousand years old—nearly immortal—then you are going to be able to produce human beings presumably who are six foot seven and have an IQ of two hundred, and, if this scene works, and you have people who are superior to others, aren't you going to have a huge caste system in this society? those who are the beneficiaries of genetic engineering and those who aren't?

STOCK: I would say that it is hard to see what sorts of limitations there will be in this realm. Because there are not obvious limitations does not mean that the battles that we fight today do not have to do with real issues that are not related to super enhancements. They simply are not. You say, "Well you don't care about embryonic stem cell work and someone like Francis Fukuyama, who has thought about this far more than you have, well Francis Fukuyama wants to actually prohibit embryonic stem cell work because he sees it as a gateway to cloning." So fear of cloning and fear of immortality and fear of these other possibilities leads to people that will make very, very damaging interventions in a process that is serving us very well and will sacrifice real people with real diseases and real pain.

MCKIBBEN: These are very unfair accusations.

STOCK: For Francis Fukuyama?

MCKIBBEN: Well, no. I'm saying that at the moment we should not do stem cell research until people have been willing to buy off on this series of things that would make it much less likely that if we move into these other realms.

KONDRACKE: Can't you draw a line between therapeutic cloning and germ line genetic engineering?

MCKIBBEN: You can. It takes work to do it and that is what we should be doing.

STOCK: OK, so here is the slippery slope that I think is a very dangerous one, and that is that you say that you do not want to do embryonic stem cell work until you're able to essentially erect a barrier between that and reproductive cloning. I would assert that you will never be able to erect that barrier, because these technologies are simply opposite sides of the same coin; they grow out of the new understandings that are emerging about life. So if you want to stop reproductive cloning, if you want to stop germ line engineering, you need to stop in vitro fertilization, you need to stop therapeutic stem cell work. In fact, I've talked with Leon Kass, and that is the direction that this will have to move if you're going to try and prevent these possibilities from emerging.

MCKIBBEN: Personally, I'd settle for a ban on germ line engineering at the moment and go from there—if everyone would sign on to that.

KONDRACKE: Which is appropriately banned by FDA regulation? What is the status of things?

MCKIBBEN: Until it's proved to be safe, it's recommended against.

STOCK: When you talk about germ line interventions, they're not banned, but you couldn't do them in a safe enough and reliable enough fashion that they would be feasible for human beings. If you knew how to do germ line engineering safely, there wouldn't be anything to do at present. So this is beyond our capacities at present, but I think it will not remain that way for very long. If you say, "Well, all I want to do is ban germ line engineering," you know that's certainly something that wouldn't be very damaging at present to the kinds of efforts that are being made in biomedical research. However, germ line engineering is going to emerge as a result of all of the germ line manipulations of mice and other lab animals, so if you really want to stop germ line manipulation in humans, rather than just pass a law that says it's illegal, you'd have to stop all of this other research as well. It's very, very difficult to draw these lines without injecting religion and philosophy and politics into just the basic discovery process.

MCKIBBEN: Injecting religion and philosophy and politics is precisely the meaning of this gathering here today; I mean if we're not going to do that then why even talk about it?

KONDRACKE: We are going to do this whole process according to our values.

STOCK: Right. What I am trying to get at is not when you talk about specific applications, that's fine. You want to ban reproductive cloning; I have no problem with that, even though I think it's a sideshow and it's not very important. The birth of the delayed identical twin is not going to bring down Western civilization when it finally is done—and it won't be done for a little while.

KONDRACKE: There is a question from the Web, and I don't want to ignore a question from the Web because we want to encourage them. This is a good question apropos of this: "Shouldn't the role of aging research be to extend the life span of long-lived people? In other words, would Bill McKibben oppose changes in aging intended to eliminate disease even if it also leads to lives of 115 or 120 years?"

MCKIBBEN: The goal of aging research that I think is legitimate is to do what some Asian researchers have called rectangularizing the curve, i.e. to move us as far as out as we can toward our 115 years of whatever it is in good health. Leonard Hayflick said the ideal would be live to 100 or so in robust health and quietly slip away. That strikes me as useful.

KONDRACKE: Before you answer, if anybody in the audience would like to ask a question there's a microphone right over there and we invite your participation.

STOCK: Two comments on that. First of all, we can talk about rectangularizing life expectancy and such—this is very abstract, this is philosophy, this is very different from when you're in the trenches and you're trying to develop an intervention for some particular disease—then you're in the forest and you only see the trees, not the forest, so you're imposing that in larger sense. That's where we differ in terms of this sort of abstract discussion of where these things might eventually lead and what the challenges would be that are associated with them.

MCKIBBEN: But you're just saying there's no point in having them because we should just go ahead?

STOCK: No, what I'm saying is there's a tendency to project our hopes and fears into the future and that when you see a person's vision of the future, it tells you an awful lot about whom that person is, about their values, and what they are afraid of. It seems to me that we really don't want to base our legislation upon such projections, because they tell us more about the person than about what the future is going to be. You're going to see that when there are concrete problems that emerge—that is the time to deal with them.

MCKIBBEN: I don't want to wait until we've engineered a generation of children. I'm not inventing my dark fantasy from some set of fears, I'm going to texts like yours, trying to enunciate what the future holds, and you say it holds parents who are engineering their children to be more religious. I think that that borders on the monstrous.

STOCK: So you've said you don't have any problems with screening of embryos?

MCKIBBEN: No I didn't say that. I said—

STOCK: As long as it is circumscribed.

KONDRACKE: Let's pause and take a question from the audience. Identify yourself, and thanks for coming to our second session.

AUDIENCE MEMBER (BAILEY): I'm Ron Bailey, the science correspondent for *Reason* magazine.

KONDRACKE: And a participant in our first discussion.

BAILEY: I'm puzzled about something that Bill McKibben seems to be worried about. You're worried about bionic children and semi-robots later into the future and there is inherent in this a notion that somehow or other randomly obtaining genes the way we currently do somehow confers freedom upon people. If that's the case, that's a lost cause because the fact is in about ten years we will know what genes we all have and it won't be a mystery. We'll know what genes give us our temperament, why they affect our intelligence, and our possibilities, and so giving those genes to people randomly doesn't confer freedom.

MCKIBBEN: That's a very interesting question, and I guess the best one could say is that the society we've constructed over a millennium operates on the notion that there's enough of fate and free will in that sort of random combination of genes to allow us to be humans as we are now. Going past that to the point where we're picking them out of a catalog and assigning them to our children with agency, instead of by somewhat of chance, creates not only an extremely different relationship between generations, but also a very different self-understanding. Imagine yourself as the child who has been programmed by the parents to be happy, to have whatever set of genes are increasing her dopamine or serotonin level. She reaches an age of sixteen and finds herself in a sort of intellectual and emotional quandary. She's happy, but is she happy because she's happy? or is she happy because she's been programmed to be happy? You could say that perhaps you'll be able to program her enough so that such bad thoughts won't arise in her head, but if you did that, and to the point that you do that, you've created something very different. We're all influenced now by our parents, but part of the act of growing up is the rebellion against that, and rebellion is not possible in the same way if every cell in your body is expressing a protein that's been selected by your parents.

STOCK: So Bill, I don't think you quite responded to that question, and that's a very crucial point. Let's take it very concretely. If we have ten embryos, and we can screen those embryos for various dispositions—because the engineering you're thinking of is the parents' engineering the child—but really it's the implications that one's genetic constitution has for who one is. Now we will know all of this information within a short period of time. If a parent is to pick one of those ten embryos, and they do it because they think there will be a predisposition to be a little happier, or whatever, now how does that child have any less freedom than if that child happened to have been born randomly? The child will know his or her genetics, his or her particular genetic constitution. You're saying that you would like to think that that genetics somehow allows this child more free will than if somebody makes a choice, but I don't see that.

MCKIBBEN: I think that on a spectrum, that is not as bad as inserting whatever artificial chromosome you're going to be able to buy from Pfizer with a particular set of them, but I think that it retains many of those same difficulties. We should not be selecting our children in the same way we select our dogs because we have a small apartment and we don't want a dog that's going to run around a lot. What's OK with dogs is not how human beings should be relating to each other.

KONDRACKE: Are there any other questions from the audience?

AUDIENCE MEMBER (SPROTT): I'm Dick Sprott, executive director of the Ellison Medical Foundation, and I think with this comment I represent about half of those who do biomedical research. Greg and I differ in our point of view of what's possible and it has some considerable impact, I think, on the direction a debate like this might go. I should tell you at the very beginning, I am not convinced at all that the topics of this debate are anywhere in our near future. I am really not convinced that aging is a disease or genetically programmed in the a that will allow us to attack it in the same way as we attack diseases. So given that premise, then I think one of the real dangers of the push in this direction in the press, and the push for specific kinds of legislation to deal with it, poses for us is that we run the danger of wasting very precious fiscal, intellectual, and political resources on a very highly unlikely goal at the cost of not pursing research on the diseases of aging that would in fact improve the lives of virtually everybody on this planet if pursued in that way. So I'd be interested in your comments.

STOCK: I would not argue with that, and I would certainly say that meaningful germ line manipulation of embryos is at least a generation or two away. First of all, it would have to compete with the genetic screening technologies that are far more effective in doing the kinds of simple manipulations that could occur, and I don't think that it relates very much to the aging field at all, which is the point that I tried to make.

SPROTT: I don't either. I really am much more concerned about spending the intellectual capital on genetic manipulation of existing humans to modify the life span—that is gene therapy if you will—to modify aging, which I think many people do think is on our horizon.

STOCK: Now as to what is actually going to emerge: I've been trying to make the point that the immediate things that are being done are the kinds that you're talking about, attempts to cure diseases, attempts to enhance individual lives as opposed to these dreams people have of immortality or of greatly extending human life span. It's only a dream that they have, maybe it's possible, but if it proves that that is fairly likely or more plausible, then we'll deal with that. To me, the fault is in debating in this abstract philosophical realm, but all of the efforts to impact the future have to come in the contact of the present day research that is going on, which is very much in the realm of biomedical research.

MCKIBBEN: The idea that people are inventing out of whole cloth some idea that germ line engineering is…to quote you, "In light of our yearning for immortality the underlying biology of aging may well be the first germ line intervention to truly tempt us." It was you that raised the specter of it, and in fact in a conference you organized on germ line engineering it was one of the repeated themes. The point I'm trying to make is, if this is true, germ line engineering isn't necessary, then let's take it off the table and let's stop talking about the inevitable redesign of human beings and the design of babies and things. Let's take that out of the equation and content ourselves with work in a much more limited and human sphere—not talk about going post-human, not talking as you do about fast-forwarding human evolution to the point where we are no longer ourselves.

STOCK: When you talk about fast-forwarding human evolution, it seems evolution goes very, very slowly—things that would occur in one hundred years or two hundred years or five hundred years—that's just an instant in the evolutionary time scale. To fast-forward evolution doesn't mean you see it tomorrow. Now, if you're talking about my belief that attempts to do antiaging research are going to be a reason that people move into germ line engineering, that does not mean that the major research in the field of antiaging medicine is going to be germ line for the foreseeable future. So those two things are rather different, and I think that was the point that was really being mentioned.

MCKIBBEN: And work that can be done without raising those kinds of problems, those kinds of possibilities, without making them any more likely, without leading us in that direction is work I think that people will—I think there will be no point in having a debate about it here.

KONDRACKE: We're going to have to stop and let each of you have a final two minutes.

STOCK: I think that we need to look at these things in a larger context, which is that all of the things we've been discussing today, in particular the most challenging ones like germ line engineering where there is a question of exactly how long it will take, or efforts to retard aging, these may bring us great difficulties, but they are on the side of life. There's no question about this. People who are envisioning these things and trying to make them happen believe that they are going to enhance human life.

I think that our dangers do not lie in this realm—the side of life—we will deal with these sorts of things. The real dangers ahead from the unraveling of human biology come from the other side, the dark side, things like the weaponization of anthrax, smallpox, or bubonic plague, all of these things that are also going to occur with absolute certainly as a result of our progress in biotechnology. I think that we tend to forget about those possibilities, and that actually the advances in biomedicine, the kinds of things that are occurring today, research that should be pushing ahead with speed and vigor, are the very developments that will ease our passage through these other challenging minefields. What we need is the development of multivalent vaccines and new ways of detecting these sorts of pathogens, new cures for these kinds of infectious diseases.

These dark agents are the things that are really going to be challenging for us, and I think that if future humans look back at our era from five hundred years in the future, they probably aren't going to see it as this horrible moment when we basically faced all these challenging possibilities and altered human biology in ways that were a disaster. Basically, they are going to look back at this moment and they are going to see it as this extraordinary moment in time when we made breakthroughs that established the basis of their lives and their vision of the future before them. And that's not just artificial intelligence and moving out into space, it is the reworking of human biology, and that is going to come gradually at first and then it's going to pick up speed. For me, it's a real privilege to be alive at this moment, when all these debates are taking place and when this is occurring. What is so extraordinary is that we are actually the architects of the changes as well as their observers. I am not ashamed of this, I am proud of it. I think it's a wonderful thing, and I think that the challenge for us is that we are also the objects of these changes, and of course, this is why Bill is so concerned.

These developments will impinge our lives, our future, our children, and our health. That is why our choices are so difficult. I think we need to be as brave as we can, and move ahead into this future trying to minimize the negative consequences and expand upon the beneficial ones. I'd like to close with a quote from Thucydides, he had a very good comment about this, one that is quite relevant today as it was in 430 B.C. He said, "The bravest are surely those who have the clearest vision of what is before them, glory and danger alike, and yet notwithstanding, go out an meet it."

MCKIBBEN: Let me say for the record that I'm against the weaponization of anthrax and the bubonic plague. I think that you've diagnosed very nicely where we need to really have this out over the next few years, and that's when you say that you're on the side of life. I hope that I am as well, and the next question is, what it does it mean to be on the side of life? I think that we need to be extremely careful and not trivialize the idea that meaning is important. I said earlier on—and I don't think you really answered—this sort of naïve consumer notion that more is always better lies behind an awful lot of this. There are times when more is better: living until eighty-five without Alzheimer's disease is better than living to eighty-five with Alzheimer's disease. But that doesn't tell you anything about living until 185 in any state, and it certainly doesn't tell you anything about living to a thousand.

We are extremely lucky to be alive at the moment; our lives are in may ways filled with ease and comfort and convenience, and we should be extraordinarily careful about trading in that world for what's behind door number two—especially if we can sense that behind there, there is perhaps some kind of vacuum, not some great adventure. Life now, life lived, life really lived as a human being, is a great adventure already and it doesn't need to be twice as long or with twice as high an IQ, or whatever, for it to be real and for it to be complete in some deep sense.

KONDRACKE: I think that's a wonderful conclusion. Thank you so much for joining us out there on the Web, and thank you for joining us in person. If this isn't profound stuff, I don't know what is. Come back next month; we're going to be having another one of these high-level debates about the human future. Thank you so much.

End.

Will Therapeutic Cloning Fail or Foster Future Aging Research?

Charles Krauthammer, Journalist
Michael West, Advanced Cell Technology
Morton Kondracke, Moderator
April 22, 2003

Pictured: Michael West, Advanced Cell Technology; Charles Krauthammer, journalist; and Morton Kondracke, moderator.

For more information on debate participants and SAGE Crossroads go to
www.sagecrossroads.net

KONDRACKE: This is a momentous month; it's the fifty-year anniversary of the announcement of the discovery of DNA, and it's the month when the completion of the Human Genome Project has been announced.

In the last two debates we have discussed questions such as: How long are we capable of extending human life? or can we make humans immortal or nearly so? Should we make super humans? Are we moving into a post-human future, etc.?

Today's issue presents a more immediate issue, but one no less profound. It's the issue of cloning human embryos for research. Our two guests are Charles Krauthammer, a critic of that research, and Dr. Michael West, whose company has actually done it. Dr. West was the founder and vice president of Geron Corporation in Menlo Park, California, and organized the scientific effort to extract stem cells from human embryos in the first instance. He is now the president, chairman, and CEO of Advance Cell Technologies of Worchester, Massachusetts, and in November 2001, he co-published the first report his company did on the cloning of a human embryo.

Charles Krauthammer is a doctor, a graduate of Harvard Medical School turned journalist, a columnist for the *Washington Post*, and a Pulitzer Prize winner. He is also a contributor to the *Weekly Standard* and the *New Republic* and a contributor to Fox News. In 2001, he was appointed to the President's Council on Bioethics.

The way we're going to do this is that we will start with Charles Krauthammer and he will do a five-minute presentation of the case against therapeutic cloning or research cloning. Then Michael West will respond or have his five-minute say. Then we'll have a question and answer session, in which I will throw questions, if it's necessary, or the two of them can interact for twenty minutes or so. Then, depending on how many questions we have from the Web audience and from our studio audience here, we'll take questions.

So, with that, let me have Charles Krauthammer start.

KRAUTHAMMER: Thank you Mort. It's a pleasure to be here. This is a very important forum and I really admire you for the work you've been doing in helping to sharpen the debate and to raise awareness about these very sophisticated, very difficult, and very critical moral dilemmas that we face in bioethics. I must say that I have to preface what I say by saying that, though I am a member of the President's Council on Bioethics, I speak only in my own capacity and not on behalf of the council.

I don't make the case against research cloning with any great relish. I understand that there is great promise with this research, although I do have some reservations about the amount of hype that has surrounded it and some of the promises being made particularly to people who are elderly and suffering about the miracle cures around the corner. I don't believe that's realistic and I believe that, in some ways, it's a cruel and false hope. Nonetheless, I understand the power and promise of this research, and it's with a heavy heart that I believe it ought to be stopped.

Let me begin, since we're going to try to do this in only a few minutes, by raising the obvious objection, which I think is one that probably influenced many members of Congress. There's a national consensus—almost a universal consensus—against human reproductive cloning. There's no dispute about that; the debate is about research cloning. Nonetheless, the research of cloning does open the door to reproductive cloning in an obvious way. Banning the production of cloned infants or babies while permitting the production of cloned embryos, I think, creates a real conundrum because if you have factories all over the country producing embryos for research or for commerce it is inevitable that someone, somewhere will implant one in a woman or perhaps, in the further future, in some artificial medium and produce a human clone, in which case, what do you do?

A law that bans reproductive cloning but permits research cloning would then make it a crime not to destroy the embryo, which is an obvious moral absurdity.

I think that is the obvious objection, and I think that it does sway many legislators, but I think it gets us off the hook philosophically. Assume that weren't an issue; I still believe that we ought to be extremely reluctant to make this legal. And the reasons, I think, are rather complicated.

There's one school, which I do not represent but which I respect, associated with the Roman Catholic Church and with others who have religious objections based on the idea of human life beginning at conception. I don't share that view, but I respect it. And I understand how, when people who have that belief, believe that research cloning which necessarily involves the creation of the clone embryo and its destruction, would be the destruction of human life. It's a serious objection, but the problem is that intellectually it's not very fruitful or productive.

Either you believe that or you don't. And if you don't believe, as I do, that human life or, at least "personhood," begins at conception, you have a more difficult problem trying to reason as to what the dangers and the reasons are that one might be extremely hesitant about legalizing this procedure. I'll take a minute to outline the three kinds of objections.

The first is what I would call the *Brave New World* objections. And that is, that by harnessing the developing human embryo, we are taking control of a mechanism of unimaginable complexity—one that, on our own, we could never have created and are going to manage. We're going to influence it; we're going to create stem cells or perhaps even organs that might be useful in therapy. Once you begin harnessing that embryo, which is this organism of incredible complexity, there are all kinds of monstrosities that become possible that are the kinds of things that all of us, I think, would object to.

We've had research with headless tadpoles and headless mice. One could imagine a situation in which one would want to create a sort of subhuman organism that could be harvested long before its birth for parts. This kind of power that we would give ourselves, I think, is very scary; it's a reason to be fairly hesitant. It is not a reason, I would say, to ban this procedure because, just as in regular stem cell research from discarded human embryos, I believe that, with regulation, we could control that and make it permissible. So it's a problem but I don't think it's a reason to ban.

The second problem I would call the "slippery slope." Once you begin to allow the production of the cloned embryo to say, the blastula stage, it will be inevitable that there will be a scientist or a researcher who will say, "If you could only allow me to develop it a little bit further, I could produce even more miraculous results."

Obviously, this would require implantation, either in a woman or an animal or, perhaps in the future, in some artificial medium. Nonetheless, instead of today, the idea of research cloning is you create a blastocyst, you tease out the stem cells, you grow them in a medium. You then tweak them so you can differentiate them into any particular kind of organ lines. You then re-implant them in the original host. That's a very Rube Goldberg way to go about stuff.

All you'll have to do is allow that embryo to grow another week or two or three, or month or two or three, and you can have a full kidney, a full lung, whatever you need for implantation. It will be a temptation extremely hard to resist. And I think that slippery slope will be out there. Today, we can say we're going to stop at a blastocyst; tomorrow who's going to be there to stop us?

This is also a serious objection. And again, I would say it's analogous to our debate on stem cells in that we've now decided that we can do that, allow the development of the blastocyst to the seventh, eighth stage, tease it out, use it and rely on regulation and the force of law and custom and the manners to actually enforce it. Again, I would say, this too is a serious objection, but perhaps one we are able to regulate and control.

So why, in the end, would I object to research cloning? Because I think it goes one step beyond normal stem cell research. It's not just the *Brave New World* factor, it's not just the slippery slope; it is that, in cloning, you do something today that you do not do in stem cell research. We are creating an embryo for the sole purpose of its destruction and use. Unlike stem cell research, in which the embryos are used, are taken from fertility clinics and are destroyed and are discarded and would be doomed anyway, here we are taking a cloned embryo, creating it entirely as a means. And I think that is breaking a moral frontier. I think once we begin to do something of that sort, which is to make the embryo entirely instrumental, entirely a means to another end, I think we have broken a barrier which is extremely important that we retain. It is to reduce the embryo to utter "thingness." I think once you do that, all the other restrictions I talked about earlier on—stem cell research and other embryonic research—are in jeopardy.

Therefore, I argue, that that would be a powerful reason to restrain ourselves, despite the benefits which might be possible, to cross a frontier which might prove to be calamitous to our moral and ethical health.

KONDRACKE: Thank you, Charles. Dr. Michael West.

WEST: Well, thank you. It's a pleasure to be here.

I'm in favor of embryonic stem cell technology and the use of cloning, if the use cloning is to clone cells and not people.

First I wanted to lay some groundwork. It's interesting that we're talking here today in a meeting sponsored in large part by the Alliance for Aging Research because these technologies, embryonic stem cells and cloning, were at least in part—a large part in fact—developed to put new tools in the toolbox of the geriatrician.

And why do I say that? Medicine today, scientific research in medicine today, is in itself a crossroads—maybe coinciding with a shift into a new century, a new millennium. In the past, medical research has focused largely on the diseases of young people. Every medical school in the United States has their department of pediatrics and obstetrics. Very few—maybe less than a handful—have a department of geriatric medicine. This is the new frontier of medicine.

Increasingly, we've solved the problems of the diseases of young people and we're seeing a rectangularization of the survivorship curve. Today, for instance, if we put all of our national resources on eliminating all death before the age of 50, the impact on life expectancy would be about three and a half years. The reason is, increasingly so, and certainly true today, that the majority of diseases we see now are age-related diseases: Alzheimer's disease, Parkinson's disease, and so on. But that has not historically been our focus.

So gerontologists such as myself have gone back to the blackboard and said, "Let's invent a whole new type of technology and call it 'regenerative medicine.' Let's go back to very primitive embryonic cells and try to find within them brand new technologies that could be used to solve these problems we've never before faced."

Now in doing so, of course, we've entered a territory of immense controversy. I think we have decades, maybe centuries, of hard ethical thinking and debate ahead of us. But I would argue that one starting point here in these first few years of this new century, one important ground rule, is that it should be a dispassionate and reasoned debate. References to "headless clones," and all due respect, *Brave New World* rhetoric are inflammatory and meant to scare people. It is not in the human best interest. If you cannot defend an argument based on intellect and reason alone, I would argue that it's not a good argument and shouldn't be used.

There certainly is power in this new technology we're developing; it could be abused. But this first argument, that this is a brave new world and we ought to be scared of it, I think is flat wrong. I think we ought to face our future with courage and a desire to help our fellow human being. And we should learn from history—history is replete with objections to the use of anesthesia in childbirth because it's against the book of Genesis. We should be against blood transfusions because the life is in the blood according to the book of Leviticus in the Bible. I think we should learn from history that we should face the future with courage and our guiding light should be compassion for our fellow human being and using our imagination and creativity to that end.

The second argument of "slippery slope"—that we cannot practically implement these new technologies because they will inevitably, unavoidably slide into an ethical morass—I find to be a bit ironic because we are within shouting distance of the U.S. Congress.

I would argue, not being a political scientist nonetheless—I think history would teach us that it is possible to put in place laws with heavy penalties that would prohibit any person in the country from crossing a particular line. In the case of the medical use of cloning, making cells not people, all you have to do is say that embryonic cells cannot develop past fourteen days. That's a limit based on biology; that is the point at which human development begins. I would argue that it would actually be relatively easy to implement such legislation, and it certainly deserves a try. And then see whether or not those penalties were sufficient to prevent an inappropriate use of this technology.

One last criticism that we heard is commodification. Aren't we headed into a world where biotechnology and businesses commodify life itself? I would argue, again taking from history, that we've heard this argument before. In the 1970s, when science wanted to help women have children through in vitro fertilization, making an embryo in a test tube, the commentaries and newspapers were replete with pictures of test tubes lined up in a laboratory with little babies in them. And the idea was that we'll have an assembly line; we're going to commodify life; it's going to be a manufacturing process. It's the same with blood banking; what could be more commercial than a bank? Some religious groups, such as the Jehovah's Witnesses, thought life was in the blood. We shouldn't be making a bank out of a fish, the human soul.

All I can say is, in retrospect, the lessons we've learned through history show that if the goal of medicine is to alleviate human suffering and the commodification or the industrialization or the making a business out of this is a means to allow such therapies to be available to the broadest possible audience at the lowest possible cost, then history teaches us that it's a good use of technology.

KONDRACKE: Thank you very much. Charles, let me just ask you—

KRAUTHAMMER: Can I respond?

KONDRACKE: Yes, I do want you to—why don't you respond and then I'll ask questions.

KRAUTHAMMER: First of all, if we're going to talk about scare tactics and sophisticated, rational debate, I would think it rather odd that you would refer to Leviticus and Genesis when I said specifically that my argument was not based either on religious or other grounds. There are people who do object on grounds of religion and I respect it, but that was not my argument at all. It is classic in this debate—and I've watched it over the last few years—that people who want to promote this brave new world of technology always like to cast the opposition as religious obscurantists. A good example is going all the way back to the example of Galileo; we're not talking about the church imposing views of people who are referring—or Jehovah's Witnesses objecting to the blood banking. It is not an argument to say that because people objected to a unreasonable advances of science in the past, therefore, the advance I'm offering today ought to be allowed. It's not an argument. You have to justify what you're doing in terms of the risks.

The two objections I raised, which I believe in—the ones that I emphasized at the beginning and the end, you have not answered. The first is you say that you can regulate fourteen days by law. Well, how are we going to regulate if we have an industry here and around the entire country of people producing embryos—cloned embryos—for research? How will you prevent a single one of those from being planted in a woman or even in other media in the future? The answer is, there's no way to prevent that. If that happens, what do you do with that cloned embryo? You're going to have the moral absurdity that says that it's going to have to be destroyed.

And the last point I brought up, which is the one I tried to emphasize, is that when you go to cloning you break new ground. It's not like stem cell research. We are using the discarded embryos—processed embryos that would otherwise die. We are saying that we are going to do something new; we're going to create embryos entirely for their destruction. You have argued that the opponents have tried to scare people by talking about slippery slopes, and you can enforce them. Let me give you an example of a slippery slope:

Two years ago we had an argument on stem cell research on the Hill. We were assured by the proponents of that bill that this meant the use of stem cells from embryos discarded from the clinics. It was emphasized by those who supported this research, including Senator Hatch and others, that they would not countenance and indeed would not allow the creation of embryos—not a clone—just regular embryos, purely for the purpose of creating embryos to be discarded and be used as stem cells. So here we are in one year arguing that we will allow discarded embryos but not the creation of embryos entirely for their destruction. A year later we have a debate on cloning and the same senators who had assured us the year before that they would explicitly outlaw the creation of embryos purely for their destruction and research are not supporting a technique—cloning of embryos—which requires the creation of embryos in order to destroy them. That is a slippery slope. It's not hypothetical. It occurred in Washington, in public, within the last two years.

KONDRACKE: Michael, do you want to respond?

WEST: I would love to.

KONDRACKE: OK.

WEST: Well, with regard to the slippery slope, the fact is that there are members of the U.S. Congress that are not scientists that are trying their hardest to understand a new technology, grapple with it, try to understand it. They then first look at the issues of the first embryonic stem cells, how they're going to be used and then, later, grapple with the issue of nuclear transfer and realize, "Geez, you know, I did begin by saying that we would never create an embryo for the use of stem cells." Then they recognize that, "Gosh, we need to take advantage of the fact that nuclear transfer offers a way of making itself that wouldn't be rejected." I don't consider that a slippery slope. I saw it happen. I saw Senator Hatch for instance, initially take that position and then I was happy to have the opportunity to sit down with him. He asked a lot of questions, and I saw in his own mind that evolution—which I didn't think was a slippery slope at all. What I observed was a collection of data, careful thought about where this technology would be taking us. Indeed, weeks and months of, as he described, "prayerful consideration." I saw what was not a slippery slope, but the use of reason and compassion about how technology could be used in medicine.

KRAUTHAMMER: Forgive me; that's not an answer. You're giving a re-definition of what occurred. You call it an "evolution"; I call it a "slippery slope." The fact is that whatever avenue he went through, whatever meetings he went through, whatever data he looked at, he made a revised moral judgment. It's not a question of fact. The judgment he made originally was, "We're at a frontier of ethical action that is very dangerous because of its inherent power." Therefore, he originally said, "Yes, we will allow discarded embryos, but we will not allow the creation of human life solely as a means to an end that violates a Kantian ideal of a categorical imperative." That was his view. Then, a year later, after consultation with you as I now understand—now I know how it happened—he says, "Oh no, the moral frontier now is breakable; I have a new frontier."

I'm not trying to say he's a bad man because he changed his view, all I'm saying is it will be tempting today and tomorrow and forever to move the moral frontier when people like you approach and say, "I can offer you some miracles." That's the danger. That's why I think we have to draw a line and stay with it; otherwise there is no frontier, no bottom and no barrier.

KONDRACKE: Michael, in one of Charles' articles, he referred to your own company developing, in fact, a fetus farm with cows. In other words, developing a cow embryo to the point where organs could be harvested. And now Charles' fear is that someday that will happen with humans. What's to prevent that if—in view of the fact that, "My child is dying…" why should that child not have a heart or a set of lungs, or something like that which we could get from this half-developed fetus? Why could that not occur ten, twenty, thirty years from now?

WEST: Well, science and technology are littered with similar examples. All power of knowledge over nature could be used for evil, and it could be used for good. And this argument that someone, certainly, will pick up this technology and use it for evil would stop all this…

KONDRACKE: By the time we ever got to it, it would be the argument that it was for good and it *would* be for good. It would have good purposes.

WEST: But Charles' criticism, which I read, and was saddened to read, was again, I would argue, not an attack based on reason but, in this case, not an argument, but an attack on our character. And what he said was, we report in the scientific journal that we had done certain experiments in animals, in this case the cow, and we cloned not just this little ball of cells, which we're calling an embryo, this microscopic ball of cells, but we actually cloned a cow fetus, a fetus with a beating heart. We took that fetus out, took tissues from it, and used it in a study of whether cloned tissues are rejecting them. And his argument was, "Look, they're already doing this in animals, so no matter what he says in pubic…"—what I say in public—"…that we should not go there with humans." It's, "You can see what they're doing with the animals and can you believe him? Is he telling you the truth?" That's a personal attack against us. I think I'm—

KONDRACKE: Frankly, I didn't read it as a personal attack. It struck me as—

KRAUTHAMMER: Where did you read that?

WEST: That was the *Time* magazine piece.

KRAUTHAMMER: And what did I say?

WEST: "Can you believe them? They say they will not go there. Can you…"

KRAUTHAMMER: I was not talking about you at all. That's not true. I have here the article I wrote and I could have Mort read it. All I said was—

KONDRACKE: Frankly, I've read practically everything Charles has said, and what he's talking about is slippery slope and he's not attacking—

WEST: It doesn't matter what this person—

KRAUTHAMMER: Well, it does because you just accused me of making ad hominem arguments, and I don't and I didn't. All I said was you had demonstrated how, in a cow, you can proceed beyond the blastocyst and produce a well-formed organ—I believe it was a kidney—whose re-implantation in the original cow had some real function. I said, "There's an astonishing development and it shows you that we will be tempted in the future to say, 'Why do the Rube Goldberg pulling out a stem cell out of blastocyst when you can just proceed for a week or two or three or a month or two or three and produce a formed organ which is far more useful and easy to use?" I'm saying that if the technology is going to be here, there'll be a temptation to use it.

And I don't know where you got this idea of ad hominem argument, but I object to it and I think it's a false accusation.

WEST: Well let's put that aside.

KRAUTHAMMER: Well, it's hard to put it aside.

KONDRACKE: Well, put it aside.

KRAUTHAMMER: All right. In deference to our chairman, I'll—

(Laughter)

WEST: But the point is this: we knew that in doing that study—and we debated this—we knew that there would be opponents who would pick up on that and say we intended to do this on humans. We don't. I'm in favor of a long—

KRAUTHAMMER: I never said that.

WEST: I wasn't speaking about you. The point is that such an experiment is ethical to do in animals. It lays the groundwork for what we call "research cloning" or "therapeutic cloning" and what we're advocating is a red herring. What we're advocating is that we clone an embryo, a microscopic ball of unformed cells that has not begun to develop yet. We propose that we set a fourteen-day limit because that's when development begins; it should be made law. We should ban the use of cloning to make people. We should open the door, as the National Academy of Sciences and many other groups have, to use this technology to make cells and, if proper force of law is placed on it, I think that we could prevent it from being abused.

And, Charles, you think the fourteen-day limit, which works in Britain—there is a law that allows a blastocyst to be developed no longer than fourteen days and, so far as I know, it has not been broken—might not hold.

KRAUTHAMMER: I'm willing to believe that it can be, which is why I support stem cell research. In that respect, as I tried to indicate in my presentation, stem cell research and cloning are identical in that respect. I'm skeptical, but because of the possible benefit, I would be willing to support stem cells from discarded embryos and have a limited number of fourteen days and hopefully it'll hold. That would be a hard, red line.

The reason why I, in the end, would come out against cloning is because of its uniqueness. It breaks a different moral barrier and that is the creation of a human embryo for the exclusive purposes of its use and destruction as a means to an end.

That I think is a different category of ethical breach, and I think that opens us up to the dehumanization of the entire process.

So my objection is not to the fourteen days. I support stem cell research; I think that we can—that we ought to at least test, but in cloning it's all or nothing; you're making something exclusively for its destruction.

WEST: This gets into your philosophical position. On the one hand you say that a human embryo is not in and of itself inviolable.

KRAUTHAMMER: Right.

WEST: But at the end of the day, you say that to create a human embryo for the sake of destruction is to breach a moral barrier. And it seems to me that you're contradicting yourself; that you really do believe deep down somewhere that a human embryo is inviolable.

KRAUTHAMMER: No. I believe that between inviolability on the one hand and "thingness" on the other, which means no respect due it, there's a territory and that's where I live, in that territory. I don't believe that an embryo is just like an appendage, all right? But I don't believe it deserves the kind of respect that a person does. I wouldn't give it a funeral, for example. I wouldn't invest it in property rights, for example. But I don't believe it's a skin cell. Once it develops it deserves a certain kind of respect. Not the respect due a full human, but not nothing. That's where I think the distinction between my and Dr. West's opinions is.

KONDRACKE: So it's the creation of the embryo for the sake of destruction that you can't abide, even if the end is the cure of people's suffering.

KRAUTHAMMER: I believe that it is an important enough moral barrier. The complete instrumentalization of a human embryo—that, I think, it is a bridge that we ought not cross.

KONDRACKE: Now, can you deny, Dr. West, that there are going to be factors, that there are going to be thousands and thousands of these embryos produced and they are going to be destroyed—and they are nascent human life? I mean, how do you breach that moral barrier that Charles is talking about?

WEST: Well, I think we see the world differently because the way we're viewing this microscopic ball of cells that we're calling an embryo—I think it's important to point out, and unfortunately the science doesn't play as important of a role in this discussion as it should, and we don't have time for a science thrust in embryology, but the origins of human life are a surprise. Remember when you first learned about how we came to be, about how sexual reproduction occurs? I was surprised. I didn't think it should work that way. In the same way, when a sperm and an egg cell unite, we have a little ball of cells that's not yet a pregnancy. But half of the time those never attach, never make a pregnancy.

What's important to point out, though, is that these cells do not begin to develop into anything until about fourteen days after the fertilization of the egg. Prior to that, what scientists say is that the embryo has not individualized; up to fourteen days, this little clump of cells could split into two and make identical twins. Indeed, that does occur up to two weeks after fertilization.

So because these cells are unformed and haven't made the first decision, which is, are we going to become one person or two?—certainly have not begun to develop—scientists say that these are blank cells, they haven't individualized. Then, from that base, I would argue that if they haven't individualized, how can we be talking about, on a rational basis, ascribing to unformed blank cells the status of a person if they're not an individual?

And so, with biology as my background, I'm saying that if we make blank cells that have not begun to develop, haven't formed a person, they're just blank cells. If we make that a bright line and say, "We will not cross it under any conditions," we're looking to entirely comprise the use of technology—far less than using tissues to help people that are sick, from cadavers, people in comas, people in motorcycle accidents—far more problematic than using blank, unformed cells.

[BREAK IN TAPE]

KONDRACKE:…as we can from cloned embryos. What's the answer to that?

WEST: You're asking me? Briefly, the cells in the human body branch out like branches of a tree. So there are some cells that begin with a fertilized egg, the trunk of the tree and the pre-plantation embryos. And then they branch out, once development begins, into all the neurons and muscle and bones and all the cells in the body.

There are cells in you and me as adults that have the ability to branch out at least with a few of the stems or branches; they're called "adult" stem cells. As far as we know today, there are no cells that are the base in the tree of cellular life. In the adult they can branch out into all of the cells of the body, as the embryonic stem cell can. So, as a researcher on embryonic stem cells, of course, I have to point out that these cells are totally potent. By definition, they can become anything in the human body, making cells for diabetes and Parkinson's and many other diseases.

But I do not point the finger at the adult stem cell researchers and diminish their work. I think it's important work, and I think it should be done in parallel. I think we should let all flowers bloom.

KONDRACKE: We have only five minutes left. Let's go to the audience for questions. If necessary, I'll repeat the questions from the Web. Any questions from the audience? Yes, sir.

AUDIENCE MEMBER: Charles has mentioned the embryo not being human, but not nothing. Could you make comparisons to the way slaves were dealt with—legally being something like three-fifths of a person at one time in our history? Or if you go to the eugenics movement, particularly in experiments done by the Nazi regime, Jews, Gypsies, blacks, many types of people were considered something less than human or something inferior. I mean would you consider your definition of embryo comparable?

KONDRACKE: Did you pick that up?

KRAUTHAMMER: You raise an interesting point: The trajectory of our history for the last centuries has been to expand the boundaries of what we call the human family, and to welcome into citizenship members of the family that had been excluded to one extent or another. Slavery is a classic example. The suppression of women is another, and I think all of us have welcomed that. Yet what we have from the bioengineering community is an attempt to move into the other direction. To try to treat certain—let us not say "persons," but "human organisms"—as less than human, or blank or non-human completely.

I think it's dangerous because there will always be the temptation to expand the non-human—what we can use—further, further, further as a way to bring cures and relieve suffering. People will say, "Well, that they're now up to fourteen days, why not twenty-one days and why not go beyond that?" So it is the movement against the trajectory of, if you like, the social compassion of the last two centuries to restrict what we call human or deserving of human respect of some sort. That's what I'm worried about.

KONDRACKE: Michael, do you want to respond to that?

WEST: Well, it's a complex question, and it's hard to answer in just a few minutes. Basically what I would argue is that, again, inflammatory rhetoric, references to Nazis, etc., should be forbidden. What we need to look at here is the moral status of the pre-implantation embryos in one hand—cells that have not yet begun to develop; that's a biological fact. On the other hand is the life of a person that we know or don't know, a person with diabetes, Parkinson's, a person who is a living, breathing human being. On the left hand a potential human being. Then again, a skin cell is potential. On the other hand, the life of an actual human being. I think that's the moral dilemma that we're in today.

KONDRACKE: Does either of you see any possibility of a middle ground here or is this an absolute barrier, the idea of research cloning?

KRAUTHAMMER: My personal view—and I speak for a small minority because the majority of those who oppose it do it on religious grounds and are rather absolutist on this, but I speak on behalf of the smaller secular constituency that opposes it on other grounds. I do think that there is something radically different about research cloning because it involves the creation for the sole purpose of destruction. When we met this problem on the President's Council on Bioethics we reached a compromise of sorts, which was to propose a moratorium instead of a ban.

Now, I'm sure that Dr. West doesn't think that's a great compromise. What it did do is allow us to look at the issue, study it to see if there might be other alternatives. As you mentioned, adult stem cells, which, even though they are not totally potent, if they are plastic enough—usable enough to treat specific illnesses—will obviate any need for cloning of embryonic cells. Because if you're going to take an adult cell out of your own marrow, the rejection problem is not there, and if it could develop into a pancreatic cell and relieve diabetes, then you've got a cure and you don't have to go this route.

So I would say a moratorium was a way in which our council found a crude compromise, and perhaps with time we'll be able to find other means of achieving the same ends that do not violate the ethical lines that I've talked about.

KONDRACKE: That gets us back to the issue of adult stem cells as the answer, but do you have any—

WEST: I don't think this is an area of our current political climate where the compromise is what we should be looking at. I think what we should be looking at is disease. We're talking about disease and therapies to alleviate human suffering. This is serious business; it isn't just, "OK, let's compromise and we'll put a moratorium in place and let thousands of people die everyday that could have potentially been cured someday because we don't know what to do."

I think we need to decide. The United States is the leading technology country in the world, and I know for a fact that the world is looking to the United States for leadership. So I would argue that we should look at this courageously. If it's wrong, then we should determine that it's wrong. If it's a good and compassionate use of technology, as many of the scientists in the United States are saying—such as the National Academy of Sciences formally recommending this—I think we need to look at this carefully and soberly and rapidly make an important decision that the rest of the world will be looking at.

KONDRACKE: Let me follow that up with a question about what's happening in the rest of the world. As of right now, what you're doing is perfectly legal as there is no national law against cloning embryos. It's up for debate in the Senate. No one knows if a moratorium will pass or if nothing will pass; and probably nothing will pass. But what's happening? Is there a lot going on in the rest of the world so that the United States might be behind in this?

WEST: I think, clearly, the United States is taking the lead in stem cell research, with the possible exception of the United Kingdom. In the United Kingdom, maybe *the* embryo research country in the world, they've already set a precedent by debating this carefully and putting in place guidelines and licensing so that the medical use of nuclear transfer cloning can be used with the proper guidelines and the use to clone people could be banned.

But the United States is right in there with the U.K., and I think we and the United Kingdom will lead the world into these new technologies.

KONDRACKE: OK, we're at the end here. Each of you has two minutes to sum up. Charles, you begin, and Michael West will have the last word since you had the first word.

KRAUTHAMMER: Dr. West has talked about the suffering of people who are dying today and I'm very sympathetic to that. He concludes from that that we have to come to a rapid conclusion. We should begin to launch this industry, which will create factories creating human embryos throughout this country in large numbers for the purpose of their destruction? I think that is a larger moral leap. I think that if his concern is with human suffering, which I share, then I don't understand why there is this desire to pour our resources into a very ethically difficult, and, I think, suspect area, when we could be pouring it into, as you indicated, a far more promising area of adult stem cells—because, as I indicated, the entire process of developing a cloned embryo into something usable is very difficult, extremely problematic, and years away. Those people Dr. West is concerned about will be dead by the time that could actually occur whereas, the adult stem cells, which we know will not be rejected, have a lot of promise and would allow us to avoid these ethical conundrums, and, I think, ethical difficulties.

My conclusion is that it is not as the proponents want us to think: obscurantist religious people on the one hand versus progress and relieving human suffering on the other. All of us are interested in relieving human suffering. The question is, do we do it at the cost of our souls?

KONDRACKE: Michael West?

WEST: Well, the use of the word soul, I think, gives me license to also refer to a biblical concept. There's a parable in the Bible, the "Talents of Gold." Two servants are given some talents of gold and have to make a profit. One servant goes out and doubles the money and gives it to his master. Another servant is given some talents of gold and, out of fear because the master is harsh, buries them in the ground and gives his master back the original talents of gold when he asks for them. In the Bible it refers to that second servant as "slothful and wicked."

Now we have to look into the future, I think, taking to mind, in part, the lessons of the past. The collective human wisdom is that fear of the future and superstitions are not good guiding lights. I would argue that knowledge and science should be our guiding light, as should compassion for our fellow human being. We ought to be willing to take the risk of making mistakes with the assumption that we're making the best efforts we can. Now, I'm not talking about a rapid decision, but a decision made thoughtfully and as rapidly as possible. I would argue that we're wicked and slothful, especially being the leading technology country in the world, if we do not think this through carefully, dispassionately, and with the benefit of our fellow human being very carefully in mind.

KONDRACKE: Michael West, Charles Krauthammer, thank you so much for doing this. This has been a challenging and exciting debate. I urge people to log onto sagecrossroads.net for continuing articles, book reviews and online discussions. Thank you very much for being with us.

End.

Remarkable Trends in Aging Interview with Dr. Richard Miller

Dr. Richard Miller, University of Michigan
Morton Kondracke, Interviewer
May 28, 2003

For more information on debate participants and SAGE Crossroads go to
www.sagecrossroads.net

KONDRACKE: Let me just start out by asking, what exactly is aging? I mean what happens? All of us are going through it: your hair is turning white; my hair is turning gray; we're getting on. What exactly is going on with our bodies and our minds?

MILLER: Well, I think it's an important question. You said it's a warm-up question, but I think it's a key issue. The reason is that, although people who are not scientists use the word "aging" more or less in a consistent way, scientists have often used it to mean whatever they're studying that they think they might get funding for.

I think a common sense definition of aging is that it's the process that just takes healthy, young adults progressively into older people who are more and more likely to get sick—who have more and more difficulty in combating whatever threats their body or the environment can produce for them. Older people are more likely to get cancer or to get the flu, or if they break a bone, they're more likely to recover slowly. Aging is the process that creates that kind of person from that healthy young adult that nearly all of us were when we were just about to enter college.

It's important to distinguish that approach from the idea of aging as merely the study of old people. For many people, if you say, "I do aging research," they kind of assume you help to figure out what an old person is like, what's wrong with the old person.

Although I'm certainly in favor of studying elderly people, I think the impact of this kind of research that will count the most is actually work that you have to do on young and middle-aged people to try to figure out why it is that they're getting old. What changes in their body and gene expression and in their hormonal systems to actually turn them into older people? That's what I would consider aging research in its most fundamental and probably its most productive sense.

KONDRACKE: How did you get into this? How did you become a gerontologist?

MILLER: Well, there are real psychological reasons and then there are ex post facto rationalizations.

KONDRACKE: Explore any and all.

MILLER: Most kids, when they are growing up, go through a phase where the idea of getting old and dying is really scary and really something they're against, and I did, too. Most people grow out of it, but I didn't. It's important—once you get to the point that you've decided that aging research is something you're interested in and something you want to pursue, you eventually recognize that it's a good thing to be doing. If you're interested in scientific mysteries—things that aren't yet solved—where people really need to use their intuition to discover what the important cracks are, aging is right up there at the top of the list, as cancer biology was fifty years ago or infectious disease was two hundred years ago.

In addition, it has, I think, going for it the advantage that if we are actually able to learn enough about aging to intervene in the human aging process, then that's going to be overwhelming in terms of public health impact, in terms of what it can allow us to do to keep people healthy. It's going to overwhelm all the other sorts of research that are going on anywhere in the world in terms of health prevention and health treatment. I think in terms of both intellectual satisfaction and its potential for doing good for people, there's nothing that can beat it.

KONDRACKE: We will explore that—the emphasis that the federal government gives in funding to aging and to diseases—but to follow up on the first question, what is aging? What exactly is the process? Is it cell death? Is it cells incapable of reproducing themselves any longer? What is the mechanism that causes aging? Do we know?

MILLER: No, we don't know. We know a good deal now about what causes cells to die, and we know a good deal about what causes cells to stop dividing. A lot of people who think they're doing aging research study those things. They're really exciting and they've made terrific progress and it's really terrific cell biology.

But aging isn't any of that. Aging doesn't happen to cells and it doesn't necessarily cause cells to die. Aging is what takes a healthy young person into an old person who's more and more vulnerable. If someone looks you straight in the eye and tells you they know what the mechanism of aging is, they are trying to kid you. No one really knows what the mechanism of aging is. There are good ideas, things that are worth pursuing, but nothing on which an honest person can state that he or she knows to be the truth.

KONDRACKE: Animal models don't tell us anything more than our own experience?

MILLER: Animal models tell us a terrific amount more than our own experience because animal models give us all sorts of experimental, as well as observational, power that you don't get just from looking at old people. Nature has done us a terrific favor by producing animals that are like us in many ways. I'm thinking of mice and rats, in particular, whose body plans and chemistry and body mechanics are extremely similar to that of humans, but are really short-lived.

So you have two systems that are very similar to one another, one of which is aging thirty times faster than the other. For scientists that's a gold mine; that really lets you test specific ideas of what might cause aging. You mentioned cell death. There's a popular idea that differences in the degrees to which specific cells in the body are more or less prone to die contribute to the aging process. But now we've got animals like mice that live three years and dogs that live twelve years, and horses that live twenty years, and people that live eighty years or so. You can test that idea to see if your ideas about cell death and how it relates to aging meet the facts on the ground. So far, none of the ideas about what causes aging have actually met those facts. There are some pretty good ones that have not yet been disproved. There is the stress resistance idea, which I hope we'll get to talk about later; I think that is likely to provide an important unifying factor.

In addition to giving us a diversity of things to play with, some of the animal models are easy to manipulate. In mice, worms, and flies we've been able to get specific information about changes that can slow aging down. That has sort of been the major breakthrough in aging research in the last ten years—the thing that really convinced those of us in the field that aging could be modified. We know it can be modified because we can do it.

KONDRACKE: We will get into some of that, but do you think aging is a single process or is it a set of multiple different processes? There are those, as I understand it, who believe in the unitary theory and the multiplex theory. Which do you believe in?

MILLER: I think it's important to understand that both viewpoints are partly correct. It is a unitary process and it is a highly complex process with multiple avenues. It's like if you ask a political scientist, "Is legislation a process?" Legislation *is* a process. We teach it in civics class. You can bring legislation to a halt by any one of a number of specific changes in who's talking to whom and what the political leadership wants to do.

Yet, to consider it only as a unitary process is clearly a mistake. There are multiple processes having to do with how the system is structured and who's having an impact on it. Studying those individual subprocesses is really important, but to lose sight of the fact that they are coordinated in a central way is also to miss the big picture.

There is a popular theory now in aging research. I think maybe seventy or eighty percent of the pros that do this work would endorse a kind of multiplicity idea—that aging of the brain and aging of the immune system and aging of the arteries, each one of them are different and complicated and interesting and that's certainly true.

But I think the greater truth is that they all take eighty years in people and three years in mice and twenty years in dogs. That's not a coincidence. All of those rates are synchronized, and those rates are very different in different species. That's the key mechanism of aging that scientists need to pay attention to.

KONDRACKE: It's clearly genetic in some sense of the word. I mean there is some kind of a program that is at work, which determines that a human being will live so long and a dog will live so long and a rat will live much less.

MILLER: You're absolutely right about that, but the question, Is aging genetic? is often a tricky one because it depends on whether you're looking across species or within a species. So clearly, the short life span of a mouse compared to the long life span of a person is because they have different genes. If we lived in a mouse's environment we would still live a lot longer than a mouse and vice versa.

But if you consider the differences among people, how much of the difference between their age range is due to genetic differences? That seems like a similar question, but it's really quite a different question because you get a different answer. The answer there might be five percent or ten percent or twenty percent, but never more than about twenty percent no matter how you slice it.

The genetic differences among people—and also among different mice and rats—that modify aging are really important to learn about because they give us clues to mechanisms and they will eventually give us clues to drugs.

KONDRACKE: From a different point of view, we have already extended the human life span by a considerable amount, from probably in the forties somewhere to eighty-something in Swedish and Japanese women, right? How did we do that and what does that tell us about how to extend the life span even further, and furthermore the healthy life span?

MILLER: Most of the change in the last couple hundred years in the average age of death and the life expectancy at birth has had nothing to do with aging research in the slightest. I can pretty confidently say that basic aging research has not yet had the slightest impact on our knowledge of disease and how to prevent it and what to do about it. It's come largely because of our ability to detect and to combat diseases of childhood, diseases around the birth period and infectious diseases that used to kill a substantial proportion of kids and young adults.

Those changes have been the major players. Changes in sanitation and health prevention practices have been important. There have also been changes in our ability to help people in their fifties, sixties and seventies who come down with a disease. Surgery and antibiotics and some classes of drugs have helped improve the chances that a sixty- or seventy-year-old is going to make it to eighty years of age.

But those don't really have anything to do with aging research. If what we really want to accomplish in aging research is to make an impact on health, what we need to do is develop a way to slow aging down. We will know that we have done that when people are in their eighties and nineties and maybe even one hundred years old and are just as healthy as today's average fifty- and sixty-year-old. It's not a matter of immortality, it's a matter of slowing down the rate at which major chronic diseases of old age actually get us.

KONDRACKE: But just as a layman in all of this, it would seem that just looking at airports, for example, the average eighty-year-old is living a life that is a lot like what the average sixty-year old lived forty years ago. They're healthier. So does the process by which those people live healthier not teach us anything about what it takes to get those people to one hundred in a healthy shape? I don't know whether it's nutrition or it's vitamins or what.

MILLER: There are very small changes in the last thirty or forty years, say, in how healthy an average eighty-year-old is. One way to look at it is to ask how long is an average community-dwelling eighty-year-old likely to live and the life expectancy of an eighty-year old in the last thirty or forty years has gone up only slightly, a year or so.

It's a mystery as to why that should be. Demographers have noticed it. They have documented it. They haven't a clue as to what has caused it. I think that it is not likely to tell us very much about how aging research works or how aging works. We can, in animal models, increase the average life span by forty, fifty, sixty percent and have been able to do so for seventy years.

KONDRACKE: Say that again. In animal models…

MILLER: We can take a mouse, where an untreated mouse is going to live about two years, and we can get it to live three years. It's been a routine matter to get that in the lab. Those three-year-old mice are extremely healthy. They have intact cognition and they are as strong in terms of their muscles, in terms of their immunology, in terms of their reflexes as your healthy middle-aged mice.

KONDRACKE: And how did you get to do that?

MILLER: Well, the first discovery of how to do that was by caloric restriction. If you take a mouse or rat and find out how much it wants to eat and you give it only sixty percent of that, you get an extremely thin, very hungry mouse that lives fifty percent longer than a mouse on a normal diet. That was the first discovery.

The second class of discoveries over the last ten years has shown that you can do the same thing by mutating any one of eight different genes. It is now possible through at least one diet—probably two, but at least one diet—and through at eight genes, to increase rodent life span by thirty, forty, fifty percent. That, to my mind, is the very strongest evidence. Although aging is a very complicated process, there are a very small number of controls, sort of rheostats, that can tune the whole system to run slower if you know where to push, which button to twist.

KONDRACKE: Well, starting with the caloric intake model, does that apply in any way to humans? I mean, how thin would you have to be to stay alive extra years?

MILLER: Well, there are two answers to that question. The easy one doesn't actually have to do with aging research, but it's been extremely well documented that most people in a Western society, where food is extremely abundant and exercise has to be looked for deliberately, become obese. When they become obese, they are more susceptible to all sorts of diseases: diabetes, heart attacks, strokes, and cancer. It's clear that for most of us, unless we are well below the average body weight, losing weight would be a good thing. It would help us avoid those diseases.

That's not something I know much about, but anyone who reads widely knows that that's true. The kind of caloric restriction that slows aging down is substantially more dramatic. Mice and rats that are on a caloric-restricted diet that will extend their life span are exceedingly thin. So, for instance, I'm five foot ten. If I were on the kind of a diet that leads to caloric restriction, I would have a body weight of 140 pounds or something like that to reach life span extension.

KONDRACKE: How much do you weigh?

MILLER: I'm not at liberty to discuss that.

KONDRACKE: You're five foot ten, and you would have to weigh 140 pounds.

MILLER: One hundred and forty pounds. So I would have to lose a great deal of weight.

KONDRACKE: We're not talking anorexia here. I'm trying to think of this in a way that everyone can sort of picture what it would take to reduce the caloric intake if you were going to really do that. So everybody would be walking around looking like an anorexic or not?

MILLER: They'd look exceptionally thin. I mean, to do this properly, you'd have to have a diet that is adequate in protein and micronutrients, and that is not the same diet that starved people in starving countries have, because there they don't get enough good nutrition. But we're talking about restriction of calories.

This is not the kind of diet that people can do voluntarily. It's been shown over and over again that, if people are given even really strong motivation to diet they can lose some weight and only a very few of them keep it off for more than a couple of years. The research on caloric restriction is not designed to prove once again to people that losing weight would be really good for their health—people know that already. What it's designed to do is teach us how aging works because it's one of the very few things that really slow aging down.

KONDRACKE: Now this, however, applies not only to rats but it applies to dogs and all other animals in between?

MILLER: It's been tried on various creatures, on rats and mice, insects, etc., and works almost all the time. It's been tried in monkeys now in three separate studies; two in Maryland and one in Wisconsin, and so far those studies look very promising. The monkeys that are on the caloric-restrictive diet in many ways, in terms of their biochemistry, resemble those of the caloric-restricted mice and rats.

The studies will have to go on another decade or so before people really know whether the diseases of aging have been slowed down in the monkeys. I'm optimistic that that will work.

KONDRACKE: OK, let's go to genetic engineering now. In rats, there are eight or so genes that exist where we understand what each of the genes do and what have we done to them?

MILLER: Well, the first publication of this kind was only five years ago, and people look at these mutations in mice and have stumbled over the fact initially that the animals lived a very long time. The reason these models are important is that it now gives us animals that we know from birth are going to be aging slowly. So we can start to ask what is different with these animals in terms or their hormones or what genes they express or their biology? Then we can ask if they have things that they share in common, because that would give us clues as to what we can try and mimic.

To give you an example, most of these mouse models and, it turns out, flies and worms that are long-lived also share in common changes in a particular hormone pathway. It's not insulin—it's not the hormone involved in diabetes, but it's a pathway like insulin; it's called insulin-like growth factor. If you change the genes that modulate the response to this hormone, then you get extended longevity.

So it opens up a whole new batch of questions: Could we do the same thing in monkeys or eventually in people by pharmacologically changing that hormonal pathway? Would it be good for us to do it, say, only in the kidney or only in the brain or only in the muscles or only in the liver? It gives us a whole series of new things to look at that have the potential for doing far better for cancer than the cancer research or for lung diseases or for heart attacks.

KONDRACKE: These changed hormone productions defeat specific diseases, or do they slow down the aging process in general?

MILLER: It's the same thing. They slow down the aging process in general and, as a very pleasant side effect, the diseases are slowed down too because aging is the main risk factor for these diseases.

KONDRACKE: Aging renders you more susceptible to disease, weakens you so that the disease can be more likely to attack you?

MILLER: Yes. In some ways, it also makes the disease stronger and more difficult to deal with in addition to weakening your defenses. For instance, a caloric-restricted mouse doesn't get breast cancer, doesn't get lung cancer, doesn't get liver cancer. But it also doesn't lose cognitive function, it doesn't lose immune function, it doesn't get cataracts, it doesn't go deaf. It does eventually do all those things; it just does them half a life span later.

KONDRACKE: It must be very irritable, though.

MILLER: Right.

KONDRACKE: So what is the implication or what is the transferability of these genetic discoveries to humans and how far away are we from being able to exploit any of that knowledge?

MILLER: The transition is that they provide us with research tools. No one is going to say that we should take people and change their genes and make them only live in that way. That's somewhere between science fiction and arrogance, and that's not what we're talking about.

But having these genetic changes available in our laboratories makes us formulate guesses as to what makes us live so much longer than dogs and less long than whales, for instance, so that we can begin to think about how to postpone not just the diseases, but all the problems of aging—the problems that don't kill us necessarily—things like arthritis and cataracts and deafness that may not be a major contribution to the life span statistics, but which make people reluctant to get old.

KONDRACKE: We have no idea how many human genes might affect the aging process, do we? If there are eight in mice, does the genome project tell us anything about that?

MILLER: Well, it's a complex question because if the question is, does a gene affect the aging process? the answer to that is probably thousands to tens of thousands. You know, some people have genes that are more likely to make them have a heart attack or become deaf or get gray hair or get colon cancer. All of those genes in some sense affect the aging process.

The genes that I'm more interested in learning about are the genetic differences between people and, say, monkeys that make people live to be eighty and monkeys live to be twenty-five or thirty. The number of genetic changes that were necessary to create a long-lived primate like us from shorter-lived primates from which we evolved, that number is not known. But I have a rational instinct to say that it's rather small. The reason that I think you can maintain that it is rather small, is that in mice, single-gene changes or dietary changes by themselves can lead to a fifty percent life span extension. So the number of things you have to do to get life span extension of a really significant kind–the number of changes you have to make—may be as little as one or two or three.

I will bet that nature, having an opportunity and the need to create a longer-lived primate like us, knew which one or two or three changes were most likely to slow maturation down so that people have enough time to teach their kids and to do what it is they need to do. They mature more slowly than other primates, and, as a side effect, the aging process was put off. Because those are probably the same genes, the ones that time maturation and that ones that, later in life, time how long it takes us to get old.

KONDRACKE: Now, there are other factors that I have seen referred to at various times. One of them is oxidation, and anti-oxidants presumably would contribute to longevity, as would some amino acid restrictions. Tell us what promises there are.

MILLER: Well, many people think that a key player in the aging process is oxidation. The body uses oxygen—we breathe in oxygen all the time—and inside the cells oxygen can be converted into a very dangerous chemical, which can render the cells sick. And that's been an attractive idea.

KONDRACKE: Which are free radicals?

MILLER: Free radicals, that's right. So it's been an attractive idea to many people that the free radical damage helps cause some of the signs and symptoms of aging. I think that's an attractive idea, although whether free radical damage is more or less important than other kinds of damage is really not understood in a serious way.

But from my perspective there are other sorts of damage to the cells, as well, that need to be explained before we understand why mice age thirty times faster than people and why dogs age five times faster than people. We're all breathing in oxygen, and we're all bathing in glucose, which can cause other kinds of damage. We're all getting mutations from gamma rays and from x-rays that can cause cancer. Why these things take seventy or eighty years to hurt people and only three years to hurt mice is the important question, and not one you solve simply by guessing that it's all to do with oxidation.

The second question you raised had to do with special diets that have low levels of a particular amino acid, and this is, I think, really exciting work. Our lab is working on that, too. We certainly don't think that starving kids or grownups of proteins is going to be an appropriate therapeutic solution. I do think that by understanding how restricting amino acid intake does slow aging down, and whether there are mechanisms there which overlap with caloric restriction, is going to point us in the right direction. In the vast array of things we could be working on, there are probably three or four avenues that we should be looking harder at and which could pay off. Finding out how different kinds of aging laboratory maneuvers come together will tell us where we need to look more intently.

KONDRACKE: What about hormone therapy, which is lately getting a bad name because of side effects that weren't anticipated? What is the promise and what are the dangers of hormone therapy as an answer here?

MILLER: Well, that's a question that's not actually about aging research very much because it's a question about the extent to which middle age and older people might be helped by giving them specific hormonal changes. It's not an area in which I'm an expert. The seminars that I've attended have suggested that it's been over-hyped to a substantial extent. There may be some older people for whom the benefits of growth hormone therapy outweigh the very substantial risks, but there hasn't really been a sufficiently long-lasting study to answer that. Some people were excited by the early publications, which said that you took some growth hormone shots and suddenly you got muscular again. That made everybody feel great, but the follow-up experiments have been much less exciting and have begun to highlight the substantial dangers.

People are making millions of dollars peddling these notions to middle-aged and older people who would like a miracle drug to make them young again. It's unscrupulous and it is very hard to be proud of.

KONDRACKE: Now, it is your contention that the federal government would be best advised to spend more money on aging research to discover what aging is and how we do affect the aging process rather than devoting money to specific diseases. Explain why you say that.

MILLER: Well, that's certainly true. It's a politically tricky thing to discuss because I certainly understand why people would be motivated—and appropriately so—to study Alzheimer's disease, Parkinson's disease, cancer, and heart attacks. These are important diseases that everyone knows about, cares about, and would like to solve. So I'm not against that kind of research in any way.

But I think people dramatically underestimate the likelihood that the solutions to these problems will emerge from studies devoted to the basic biology of aging. To give you an example, Jay Olshansky, who's a demographer in Chicago, has calculated that if there were no cancer at all—we snap our fingers and suddenly no one over the age of fifty ever gets cancer again—that the average white American woman, who now lives about eighty years, would gain only two and one half extra years of life. If fact, Jay has calculated that if there were no cancer, no strokes, no heart disease, and no kidney disease, that this woman would gain more than two and a half years of life; she'd gain eight or ten years of life.

KONDRACKE: Then she would die of what?

MILLER: She would die of the same thing she would have died of at eighty, but now she'd be ninety.

KONDRACKE: But not one of those diseases.

MILLER: She would die of the next disease coming up, of one of the four or five next diseases that include Alzheimer's disease, getting hit by a bus and having a hip fracture, and getting an infection. She would die of something else.

KONDRACKE: That's true.

MILLER: But aging research in the laboratory, not in people but in the laboratory, has been able to do three times better than that and has been able to do that for the last sixty or seventy years. So if we've got something in a laboratory animal that works, and already can accomplish an extension of healthy life span—vigorous both physically and cognitively—that is dramatically better than the best we could hope for by conquering the four major diseases. I think it deserves a lot more attention, and a lot more brainpower, and a lot more money than it's currently getting.

KONDRACKE: Part of the politics of this must be that lots of very young people die of cancer, and that young people die of kidney disease and so on. As opposed to extending life span for older people, we want to save people who are younger, right? The whole purpose of medical research should not be simply to extend the life span of older people, but rather to accomplish a healthy life for younger people.

MILLER: You know, I don't think I agree with you there. Many of the diseases that are receiving a lot attention in the scientific community, in the public community, and in legislature as well are diseases that are scourges of people who are middle-aged and older. These people are getting Alzheimer's disease, Parkinson's disease, lung cancer, and breast cancer—all of which are diseases that are really pretty rare in the twenty to thirty year age group. They are receiving a great deal of attention, as they deserve to.

I think that people are motivated to do research to increase the likelihood that not just twenty-year-olds will have long, healthy lives but that will help combat the much more common diseases that afflict the majority of us that will make it into our fifties and sixties and seventies.

KONDRACKE: Let me just remind the audience that we are eager for your questions; I can't think of them all. So by all means, e-mail in, or those of you who are here in the audience, scribble off a question. We have received a few off the Web, and I will pose them in a couple minutes.

MILLER: Let me say one more thing if it's OK.

KONDRACKE: Yes, go ahead.

MILLER: It's in response to that last question of whether the motivation for aging research is under-funded and under-appreciated. If I were to come before an influential person, a Congressman or someone who is in charge of the National Institute on Aging, and say, "I've got a cure for breast cancer. So far it only works in animals, but I'd really like to pursue it and can reduce the chances by ninety percent that an older individual will get breast cancer." Everyone would say, "Well that's great. Let's work on it."

If I were to say, "It's got a side effect; it also stops people from becoming blind. Is that OK? Can I still work on it?" "Sure. That's a good side effect." And if I were to mention, "It also stops them from getting lung cancer and colon cancer, and they don't lose cognitive function and they don't lose hearing. Is that a good thing to work on?"

People would say yes until they understand that it's aging research that you're talking about. You've tricked them! What you really want to work on is aging research. It's at that point the doors get closed because everybody knows that aging research is not going to be productive.

KONDRACKE: Why do they think it's not going to be productive?

MILLER: Well, for a variety of reasons. Some people associate aging research with hucksters; people who are pushing growth hormone and DHEA and melatonin. People are so—appropriately—frightened of getting old that they are willing to turn to unproven remedies that are introduced to them by people with a friendly smile and a testimonial or two. The far smaller number of us who are honest and serious students of aging are sort of drowned in that societal understanding that people talking about aging are quacks or crackpots.

If Nixon goes on TV and says, "We will cure cancer," he gets applause. You know, everybody thinks that's a great thing. If a president should be so foolish as to go on TV and say, "I'm going to invest the resources of my administration to slowing the aging process by fifty percent," he or she is going to be jeered at and laughed at. That's a matter of public misunderstanding; that's a matter of society and people who run the press. The intellectual leaders who set the tone for discussion in this country have set up that situation so that's a laughable statement.

KONDRACKE: How would you reverse that? If a politician said, "I intend to extend the human life span and to help people live happier, longer lives or healthier, longer lives," I would think that that would politically be a winner. Would it not?

MILLER: I would think that, too, but you and I are in the minority. As you can imagine, I'm not going to mention any specific presidents, but individuals who say, "We're going to devote the resources of this administration to slowing the aging process down," they would be considered weirdos.

How do you reverse it? I think you reverse it by having Webcasts of this sort compulsory in all schools throughout the country for the next 20 years.

KONDRACKE: Support your local gerontologist.

MILLER: Yeah.

KONDRACKE: You would describe something as gerontologiphobia. What's that?

MILLER: My wife is an English professor and we sat down to dinner one day with a distinguished colleague of hers, a professor of philosophy. She was really eager for me to get to know this guy and for us to become friends. He was really going to be an intellectual resource for us in Ann Arbor. I told him I did aging research and he said, "No; that's not a good thing." And I said, "Yes it is." And he said, "No, that's a terrible thing to do. We don't want the world to fill up with old people. We've got to stop aging research, because if, God forbid, you should get it to work, then everybody will just become old and we don't want to be surrounded by old folks all the time."

And I tried, politely at first, before he stormed away from the table, to say that that kind of reasoning doesn't make any sense as far as I can tell. If you believe that's the case, then you ought to be against cancer research—and you ought to be against research in cardiovascular diseases. And he was in favor, as most of us are, of cancer and cardiovascular research. He's against giving away cigarettes in schools. He's against taking seat belts out of cars because, in general, like all of us, he is in favor of things that will increase the likelihood that we'll get to be seventy or eighty and be in terrific health. And I said that's what I was doing and he said, "No, no. You're doing aging research. You want us to live forever and the world will fill up with old people." He wasn't able to recognize that the good kind of research, the kind that everybody except me is doing, and the bad kind that us gerontologists are doing, were aimed at the same kind of problems and that, if successful, would produce the same kinds of situations and deserve the same kind of moral support.

But the position that this fellow took, before leaving our lives forever, is extremely common.

KONDRACKE: Right after dinner, I take it.

MILLER: Right after dinner. I was polite over the appetizers but when the main course arrived we were yelling at each other. But it's extremely common, and if I give a lecture that has an audience that is non-scientists, educated people who work in other professions, this is almost always a question that I get, "Aren't you doing a bad thing by trying to slow aging down?"

I think that's simply a mistake, but sometimes it's a mistake that people are very slow to come around to understanding.

KONDRACKE: We have a question from the audience: "What is the most important finding in aging research over the last ten years and what is likely to be the most important over the next ten years?" Good question.

MILLER: If you had asked me that question ten years ago I would have said, "Caloric restriction!" Caloric restriction is the breakthrough that proved you could slow aging down.

But now I think the most important finding is actually an observation from the study of mutants in worms, of all things. People have made a lot of different types of mutations in worms to slow life span down, and the really cool thing is that these mutations have a side effect and that is that they make the worms resistant to all sorts of stresses. I don't mean worry about the next exam. I mean if you irradiate them or if you give them an oxidizing chemical or if you heat them up a little too much—these are really bad things and worms tend to die—but the mutants that would live a long time are resistant to all of these stresses.

And the reason that's an important finding is that it's a big clue as to what it takes to expand life span and to slow aging down. It suggests that you've got to find ways of increasing the resistance of cells throughout the body to all sorts of things.

So the next step in that, which is just now becoming clear, is to ask whether stress resistance of that kind also applies to mammals. Several labs, mine included, have begun to document that if you have mutations that make mice long-lived, that these cells from these mice are also stress-resistant. So that opens an important therapeutic approach. If you could figure out what it takes to make cells stress-resistant, this might be the same thing that it takes to slow aging down. We can now study people that are going to be very long-lived and people that, like most of us, are going to die in their seventies and eighties and ask, Do they differ in their cells which are stress-resistant? Are there genetic factors? Can we mimic that with specific hormones or nutrients or behavior modification?

These are the avenues toward prevention that are very likely, if pursued appropriately, to give us ways to prevent not just one disease, but all the diseases of aging together. That's the breakthrough that I think is most exciting and deserves to be focused on the most.

KONDRACKE: But what you are doing is studying the consequences of the mutation of genes, as opposed to how to manipulate genes in humans. In previous SAGE Crossroads debates we have discussed germ line engineering and actually reengineering human beings. That's not where you're going. You're going to medicines that people would take to reduce the oxidation or stress or increase stress resistance that you learned from your genetic studies.

MILLER: That's exactly right. The genetic changes are changes that give us research tools to ask questions about the physiology and the cell biology so that we can come up with medical strategies—presumably drugs, but who knows—to accomplish the same thing that nature does by genetic changes.

KONDRACKE: OK. Another question: "Is your goal to understand the aging process for slight improvements of quality of life, still allowing for a natural course of aging or for drastic extension of the number of years lived?"

MILLER: I guess it depends on what you mean by drastic.

KONDRACKE: Is there in aging research a thought about whether there is a natural limit to human life span?

MILLER: Sure. It used to be when people were asked this question to try to pick a number that was large enough so that they'd be the first person mentioned in the lead article in *LIFE* magazine. There would be a competition to see who would push that limit the furthest.

But I think there's actual data now and there's a good clear answer: Caloric restriction extends life by about fifty percent. Each of these genetic mutations extends life span by about thirty or forty or fifty percent. So we can, without stretching a point, understand that we've got good reason that we can slow down aging and, in that way, get a fifty percent increase in age extension so that the average person who is ninety years old would be roughly as healthy or unhealthy in a variety of ways, productive or unproductive as today's sixty-year-old. I think that would be an important and exciting thing to do.

Whether one considers that a drastic or a small change, that depends on your perspective. From the perspective of a mouse, that would mean getting to live as long as a dog would, and that would be a dramatic extension of life span. From our perspective, it's not that impressive.

KONDRACKE: Doing the math here, you are saying that the average human life span could be extended by forty years.

MILLER: Yeah.

KONDRACKE: So we could live to be 120.

MILLER: If we were able to accomplish in people what is now routine in mice, it would be an extension of about forty percent—about thirty or thirty-five years would be reasonable to expect. More than that is science fiction. It's guesswork. You can't say it could never happen, but there's no reason to expect it either.

KONDRACKE: Another question: "Is gestation length somehow related to longevity? Or is there a constant ratio?"

MILLER: There are differences among different species that parallel, more or less, the differences among species of life span, and it has to do with evolutionary principles. There are some kinds of animals, like mice, who live in a very risky environment where they're going to freeze to death or they're going to get eaten by a fox. For them it's really important to have a short gestation period, get their babies out there before they themselves get eaten. And that kind of environment doesn't produce slow aging, it produces really fast aging.

Individuals like us and elephants and whales, where we're much less likely to starve to death or get eaten by a fox, have a life span that's a lot longer, and we can also afford to have longer gestational periods. So the two tend to go hand in hand. Whether it's cause and effect, which is, in some sense, an implication of the question, is a much trickier thing to decide.

KONDRACKE: OK. Looking at hormones in aging—insulin, IGF1, growth hormones and sex hormones—what are we learning about their impact on aging, both pro and con?

MILLER: Well, we're learning very different things. Insulin is a critical player in diabetes and diabetes is a scourge of all Westernized countries. Learning more about diabetes and how to prevent it and the role of insulin is a great thing to do. Whether it has much to do with aging, I think, is less likely. IGF1 is the thing that is going to be more informative than insulin research, and the reason is that nearly all the mutations in the worms that extended life span block IGF1.

KONDRACKE: IGF1 is what?

MILLER: It stands for Insulin-like Growth Factor 1. It's related to insulin but it has a very different role. Of the mutations in mice that extend life span, five of the eight also interfere with IGF1. And, if that's a coincidence, that's a huge coincidence.

So people have begun to think—and it's very sensible—that IGF1, among its other roles, makes us grow fast, is good for bones and muscles, but also may be that it times the aging process. Maybe, when we are children, it sets our whole life span so that we are more likely or less likely to make it into our eighties in good health.

KONDRACKE: So where is IGF1 on the therapeutic process? I mean, is there a pill on the horizon or injections of IGF1?

MILLER: There's no reason to think that would be a good thing. There is some reason to think that might be a bad thing, particularly for children. The genes that lower IGF1 levels in early life tend to promote life span—exceptionally good life span. But no one would suggest that you should take a batch of kids, as an experiment, and shoot them up with something that's going to make them really short in the hopes that they would be extremely long-lived.

KONDRACKE: That would be the consequence of…?

MILLER: That's a plausible guess, but no one really knows.

KONDRACKE: I see. Well, are there biomarkers for aging and how would we develop them?

MILLER: Well, a biomarker, as I would use the term, is something you can test how old someone is in a biological sense. You might have someone who's fifty years old by her birth certificate, but who you think might have the health of a seventy-year-old or a forty-year-old. You'd want to be able to do a test to see among people who are in middle age who are looking older or looking younger from a variety of perspectives.

Studying aging without biomarkers is like trying to study blood pressure without a blood pressure cuff, or fever without a thermometer, in the sense that biomarkers would provide a scale for aging. And unfortunately we don't have any good ones yet. We have a few that sort of work in mice, but the kinds of research to measure the aging rate in people is very unpopular, very under-funded. It's not exciting science, it just happens to be very important science.

KONDRACKE: This next question relates to the phobia of gerontology, and it goes to the question of evolutionary biology. One of the theories on why people age is that evolutionary biologists used to think, as I gather, and no longer do, is that Mother Nature really did want to clear the field for younger, reproductive-age humans or animals and clear out the clutter of old people. Now, I gather that that is no longer accepted among evolutionary biologists. Can you explain why?

MILLER: Well, it's a popular thing. Whenever I go to talk to medical students, there's always somebody in the classroom that says, "Yeah, it's got to be just like you said: You have to have a gene to clear out the old folks so that the young folks can inherit the farm," or something. The problem is that it's a misunderstanding of how selection works. If you actually had a gene that caused aging, that gene would go away because aging is bad for you. It doesn't help you leave additional children to get old and die. The genes that are good for you are the ones that slow aging down.

So there may be genes for slowing aging down, but genes whose function is to cause the signs and symptoms of aging are wiped out by ordinary Darwinian pressures. Just like there's no gene for making you run slow or think slow or not know how to talk, there are no genes for causing aging. They're really bad things to have.

KONDRACKE: And yet aging is such a constant in life and in biology that you would think that it does serve some sort of evolutionary function, wouldn't you?

MILLER: No, you wouldn't. Aging is really common because there's not enough selective pressure to put it off forever. A mouse in the wild lives six months. You don't need to make a mouse that'll live ten years because it's going to get eaten. You don't need to make people that are going to last three hundred years because when we were living in caves and growing up as Neanderthals, most of us didn't make it past thirty, so there was no need to design a person that could last five hundred years. People just didn't make it regardless of whether they aged or not.

Nature has evolutionary pressures to make individuals that last long enough to make a couple of babies, enough to repopulate the species, and not much longer than that. Any additional work done in that area would be wasted effort—waste of effort that could go into reproduction, which is what selection actually pays attention to. There are a lot of things, including aging, that are in nature, but are not the positive product of the selective process.

KONDRACKE: We're almost to the end here. If you were to have a message for Congress, you would tell them what as to the kind of funding levels that the aging field could absorb? And what kind of results could you produce for them, do you think, say, over the next decade?

MILLER: I think it's rash to make promises; it can lead to overly high expectations.

I think that the promise of aging research is so dramatic that the pittance that it receives is a national shame. And if the amount of funds going into aging research were to increase gradually, but dramatically, over the next five or ten years, it would have the advantage of attracting the best brains in the next generation and giving them the money they need to test out all these exciting theories.

Twenty years ago, aging research was a backwater. The only kind of research people could do was to take some old animals and some young ones and see how they differed. But now the scientific community has brought out a dozen really exciting ideas about how aging might work. There are actual exciting problems and useful methods that could be brought to bear on testing those ideas.

They won't get tested if the field is under-funded because the money is necessary, and, just as important, smart minds are necessary. Scientists, like all of us, want to have jobs, they want to be able to support their families, and they want to go into areas that are well funded.

So new funding is important to boost the community. And that's a job that scientists by themselves are not going to be able to undertake. It requires, more than anything else, the support of individuals who have the ear of the public and the ear of those who set science policy.

KONDRACKE: Dr. Richard Miller, thank you so much for being with us.

End.

Is Politics Stifling One of the Most Promising Avenues of Research? An Interview with Stephen Hall

Stephen Hall, Author
Morton Kondracke, Interviewer
August 12, 2003

For more information on debate participants and SAGE Crossroads go to
www.sagecrossroads.net

KONDRACKE: I am happy to be here again, and I am happy especially to be interviewing Stephen Hall, who is the author of this book, *Merchants of Immortality: Chasing the Dream of Human Life Extension*, which I heartily recommend to everyone who is concerned with aging research. I can't think of a book that more fully portrays what goes on in aging research, both the good and the bad, the politics, the science, the personalities, the corporate wars—it covers it all. It is a magnificently readable book. It's accessible to any lay reader, and I am sure that all the experts in the field have been eating it up, looking through the index looking for their names, as well!

But it is magnificently written, and it's a deeply moral book, as well. It points in the direction of saving lives, and it argues for the freedom on the part of scientists to do it—without being didactic, without being lecturely. I cannot praise this book highly enough.

This is Stephen's—how many books have you written?

HALL: Fourth.

KONDRACKE: Fourth. The previous ones were?

HALL: *Invisible Frontiers* kind of chronicled how biotechnology got started. *Mapping the Next Millennium* was kind of a post-modern atlas of new discoveries and how they can be mapped. And the one preceding this was called, *A Commotion in the Blood*, which is about immunology and cancer.

KONDRACKE: He has written for the *New York Times Magazine* and numerous other publications.

I should say that this event today is almost on the second anniversary of President Bush's August 9, 2001, decision on stem cell research, which restricts federal funding of stem cell research, embryonic stem cell research, to those lines that existed at the time. We will get to that shortly.

But first, let me just sort of run through the book with you.

HALL: OK.

KONDRACKE: One of the first personalities that we meet in this book is Leonard Hayflick, whose idea is the Hayflick limit. Explain who Leonard Hayflick is, and what the Hayflick limit is.

HALL: Leonard Hayflick was a scientist who was then working at the Wistar Institute in Philadelphia. This was in the early 1960s. The Wistar Institute was very heavily involved in vaccine preparation in those days, and he was assigned this task to try to grow cell lines that vaccine viruses could be grown in to create safe vaccines.

In the course of doing this, he grew a cell line that was derived from fetal material that had been sent to him from Sweden. He plated these cells out and he watched them grow for about eight or nine months, and then, all of a sudden, they stopped growing. And he thought, like most cell biologists in those days, that he had done something wrong—either he hadn't fed them well, or he hadn't tended them well—he had done something to mistreat his cells and they didn't continue to grow. At that point everyone thought that cells were immortal, that they would just keep going on and replicating forever.

So like a good scientist, he did two things: he repeated his experiments and they still stopped growing; and then he'd listen to his data, even though it contradicted what the general dogma suggested at the time.

What ultimately came out of that was the discovery that cells, when they are grown in a petri dish and culture, reach a limit called the Hayflick limit, after which they stop replicating. They don't instantly die, but they kind of enter a stage known as cell senescence, and that ultimately leads to cell death.

But this notion that cells were not immortal, but that they could continue to grow to a certain point and then hit a wall and then just stop, had enormous repercussions for aging research, precisely because if cells were mortal, then understanding the mechanism that caused that mortality might suggest ways of getting toward aging as a biological phenomenon and what medicine might do to treat it.

KONDRACKE: So after all the various kinds of research that has been done, has the Hayflick limit been overcome? Have researchers discovered how to keep cells alive beyond their natural limit?

HALL: I was going to say, in the dish, they have. But it took about almost thirty years for them to do so.

The Hayflick limit ultimately led to a field of research in senescence called telomere biology. It has to do with the ends of chromosomes, as many of your listeners probably know. These telomeres grow shorter with each cell division of a cell's life. At a certain point they simply become a little disaggregated, and the metaphor that is often used, and I think it's a good one, is that they are kind of like the plastic cap on the end of your shoe laces: that becomes removed, and then the end gets a little bit frazzled, and then all of a sudden the thread falls apart.

Well, if the thread is your genetic material and it's falling apart, you are in trouble, the cell is in trouble, and possibly the organism is in trouble. What happened was, researchers discovered that there is an enzyme called telomerase, which in rare instances actually adds a little bit of the telomere back on and preserves the integrity of the end. And then when some cells were treated with telomeres, they actually went through the Hayflick limit and continued to replicate with the addition of this enzyme.

KONDRACKE: Telomeres are one of the subjects of this book, and we will get into that in a second.

But all of this research has led people to talk about fountain of youth genes and immortality enzymes, etc., and telomeres would be the immortality enzyme. But Hayflick is against talk about immortality, right?

HALL: Hayflick was a skeptic, you could say. His belief is that there is really no genetic way that one could extend life span easily on the basis of the information he has seen up to this point.

It's interesting. I think he represents an earlier generation, and I don't think is quite as enamored of the use of model organisms, for example, to make genetic discoveries that have application to human beings. So I think he's at least skeptical that anything could be done in this particular area.

KONDRACKE: What about with telomeres?

HALL: With telomeres, or with the genes, actually, that extend the life span even of simple organisms. He's become a rather vocal skeptic about this. I think it is an important voice to pay attention to because he's got a great store of knowledge about aging.

On the other hand, I think this field has kind of evolved in a molecular direction in some interesting ways that bear watching, and may ultimately suggest that maybe that's not entirely true.

KONDRACKE: OK. So if I correctly understand the areas of research that are active in aging research, telomeres would be one. Stem cell therapy would be another, which is another major focus of your book. Genetic therapies would be another and calorie depravation is, as you say, the best. If you are going to want to extend the life of an organism, depriving them of enough food seems to be the most dependable, but you don't spend a lot of time talking about that.

HALL: I don't talk a lot about caloric restriction, and it is not necessarily the best—it's the only one that's been repeatedly established in experiments. If you limit the amount of caloric intake, the amount of food that these model organisms ingest, they very predictably and reproducibly live longer.

You can get back to this later, but it actually kind of curls back into some of the more recent research in a very interesting way.

To take a slightly longer view, what I am talking about in the book are kind of two different general areas of research. I think of regenerative medicine, which would include stem cells, as a kind of continuation or part of the continuum of medical progress that has been made over the last century—part of antibiotics, part of vaccination—it's all part of extending life through medical intervention. It doesn't necessarily extend life span, which is what possibly some of these immortalizing enzymes, or, I think, more likely, some of these genetic interventions might have.

But they are two different fields, and they are kind of converging at the same time, and they are also drawing out.

It has been a very impressive century, as I am sure, again, everybody is aware—the extension of average human life expectancy from one hundred years ago was slightly less than fifty years and now is very close to eighty. That's an incredible achievement over the span of a single century.

Much of it has to do with socioeconomic factors and people living better and eating better, but a lot of it has to do with medical intervention. I know some demographers believe that there is no reason to believe that that trend will not continue in the foreseeable future, and I think it is things like stem cells, if that technology becomes refined, that will be contributing to that.

KONDRACKE: Right. So there's a range of potential life expectancies within the next fifty years. Hayflick is saying that if we cured—I believe he's the one who says this—that if we cured Alzheimer's, Parkinson's, cancer, diabetes, all the rest of that, that we add only fifteen years to the average life span, and people would be living, on average, close to one hundred then.

HALL: That's right—about ninety-five to one hundred, which would be significant.

KONDRACKE: Right. But there are some people who will go up to 150?

HALL: Those of you who read the paper this morning saw someone who went up to 5,000 years by the end of the current century, which I think is a somewhat exaggerated claim!

But yes, you do hear these numbers thrown around. I think there are a couple issues I'd like to address that this brings up.

One is, people talk about how much longer you are going to live, but you don't hear the companion piece to that, which is how well are you going to live when you are living longer.

KONRACKE: Right.

HALL: But if the quality of life is not similarly maintained while life is being extended—and many of the elderly people who have attended some of these talks that I have given in the past couple of months are the first to point this out; they basically say, "What's the bargain?"

If you are going to be living longer and you are not going to be having the quality of life that you would like to optimally have with that extra time, it almost becomes a kind of purgatory of extended suffering and incapacity and a burden on families and loved ones.

I think people need to think a little bit more broadly about this than simply how many more years they are going to live.

KONDRACKE: Right. Now, this is an aside, which I don't remember being covered in your book—but on this point, women are now living about eighty years.

HALL: Yes.

KONDRACKE: So if they go to eighty-six, or ninety, or something like that, on average, are they living healthier lives in those added years or not?

HALL: In some respects, yes. You mentioned Parkinson's and Alzheimer's and cancer. While we are making inroads, we still have not grabbed the brass ring in any of those cases. If you could physically extend life to an average life span of, let's say, 120, in which people had typical physical vigor, and if you haven't solved the problem of Alzheimer's, you are merely creating a larger population for that disease to prey upon.

So it's not simply a matter of extending life, but it's simultaneously addressing some of these other issues. Because if you don't do that, you're not going to be able to bring everybody along with the quality of life that you would like to have.

KONDRACKE: Right. Then they've got to figure out what they are going to do with all those added years. Society is going to have to figure out how to either give them something useful to do, or how can we all afford to pay Social Security? That's another subject for another time.

HALL: OK.

KONDRACKE: Besides Leonard Hayflick, another graphic character, or maybe the most graphic character in this book, the ultimate merchant of immortality, is Michael West—who we had on a debate with Charles Krauthammer about stem cell research, and it's as close as one of these exchanges has comes to blows, actually! I sort of had to part them, because they really got into it.

But tell us about Michael West and what kind of a character he is and what being a merchant of immortality is all about.

HALL: Well, Michael West grew up in the Midwest. He actually studied psychology in college. He went back to Michigan, where he was from, to run the family business, which was a motor leasing business, when his father became ill.

The business was sold; he inherited a substantial amount of money—the amount of which he has never revealed—that clearly provided him with the funds to kind of underwrite this kind of mythological and philosophical quest about the meaning of life, which he then engaged in for about ten years.

During that period he was an actual card-carrying, I guess you could say, creationist. He believed the Biblical version of creation, which gets into the whole Adam and Eve story and the expulsion from the Garden of Eden, and why humans were going to be denied immortality because of their sins of knowledge.

With this background, he went out to these creation institutes on the West Coast and hung out with those people and picketed abortion clinics as part of that ethos.

Then he kind of convinced himself that the Biblical version of creation was incorrect and that archeology contradicted it too strongly, and he decided to become a scientist.

So from this very strong creationist background, you have a guy flipping all the way over into a philosophical infatuation, I think, with science. He attended Baylor University, he got a Ph.D. in biology, and as he was getting his Ph.D., and starting to go to medical school, the telomere story was beginning to break. He's always had this infatuation and obsession, I think it's fair to say, with aging, trying to do something to arrest the process of aging, and looking at a biological means to do so.

He, therefore, became involved in the foundation of the first company that was devoted to this area of research, which is called Geron, and was founded in the early 1990s. It was largely, although not exclusively, founded on telomere biology. There was a meeting for venture capitalists in November of 1991, I believe it was—they routinely have these things where various entrepreneurs come in, and they pitch a biotech idea and people either give a thumbs up or thumbs down!

But Mike West gave a talk about this and said that the science was approaching the point where it was actually worth thinking about commercialization. Venture capitalists were writing checks in the breakout room after this presentation.

It captured the moment perfectly. People were ready to hear this. It intersected with the interests of venture capitalists to get into an interesting area of biotechnology. It intersected sociologically with the culture that's with the baby boomers as they grow older, and is very focused on this issue confronting the mortality of their parents, and ultimately their own.

It was one of those moments where everything kind of coalesced, and this company was formed rather rapidly.

They went on to do a lot of work in telomere biology. Interestingly, it looks now like the most promising application of telomere biology may be as a potential cancer therapy, because cancer cells use telomeres.

Let me back up just a second. There really are several classes of immortal cells. We talked about no cells being immortal, but there are a couple of exceptions to the Hayflick limit, one of which are stem cells, which we will talk about in a little bit. The other are cancer cells. Cancer cells override this limit on replication and that's what a tumor is essentially is—a bunch of cells that have overridden the governor of cell replication and are replicating without cease. They use telomeres to do that. So the notion is that you might be able to inhibit telomeres in cancer cells, and that it might be a cancer therapy. So that research has kind of steered off in that direction.

I should mention that the first generation of scientists who participated in the telomere work at Geron, some of them have gone on to another company—a number of companies, actually, but one of them is called Sierra Sciences. They have become involved in the notion of trying to transiently, or for a short period of time, activate the human telomeres gene in a way that it might rejuvenate the cells and rejuvenate the telomeres in one's body and then, just as quickly, shut it off so that it doesn't trip it into a kind of cancer scenario. Now, ideally, they would like to do that in the shape of a pill, and they are doing experiments on that. I think it is very early for that, and one can conjecture that there would be certain problems with doing a kind of systemic treatment with something like that. Nonetheless, that's one avenue that is being pursued at the current time.

KONDRACKE: OK. And Michael West, then, gets into stem cells in a big way, and he is an agent provocateur of all kinds of political turmoil connected with largely his—I think you would say—exaggeration? Is that a fair expression of the way or hyping of the possibilities at every stage, making himself—

HALL: I think it is fair to say, and certainly Mike West is not the only one guilty of this. There is a kind of phase shift, as scientists would say, where scientists are working in an academic lab and then they become affiliated with the company. The phase shift involves a kind of, I wouldn't say surrender of caution, but the language in which their pronouncements are couched suddenly—the normal caveats are shed.

KONDRACKE: They want money. They want capital, right?

HALL: There are all sorts of different pressures on a public corporation or any kind of commercial enterprise that are not necessarily always present in an academic lab. They have their issues, too.

Nonetheless, there is an interest in keeping a company in the public eye, to making it sound to investors that things are moving along in terms of product development, that the promise of these products is imminent, and it's always surprising to me (a) how far out onto a limb people are willing to go, and (b) how many other people are willing to believe it when they do go out on a limb!

KONDRACKE: And journalists are only too happy to join in the fun because they can get their story on the front page.

HALL: Exactly. And we are the go-betweens. I can guarantee you that when Geron started referring to telomeres as the immortalizing enzyme, stories that might have been on the inside of the paper started appearing on the front page of the paper, because the word "immortalizing" gave it a whole different resonance in the general society.

KONDRACKE: OK. Let's go to stem cells, which is the most politically and morally controversial aspect of this kind of research.

Now, there are two fundamental kinds of stem cells that we are dealing with: adult stem cells and embryonic stem cells.

It's fair to say that the Bush administration has a bias toward adult stem cells, and the Catholic church and other right-to-life movements believe that adult stem cells, drawn from bone marrow, blood, umbilical fluid, fat, you name it, can be somehow tweaked into doing everything that embryonic stem cells can do.

Now in the book, you cite a number of examples where adult stem cells really have done some fairly phenomenal things—although the scientific community says, "No, no, no, no. Embryonic stem cells are the way to go. They are the big thing. We are going to save lives with embryonic stem cells." They tend to push adult stem cells to the side. But what's your judgment based on all the interviewing that you have done about the potential for adult stem cells?

HALL: I think that the potential for adult stem cells, and I think it is pretty clear in the book, could be quite significant. Part of the reason for that is we actually have had a great deal of medical experience with adult stem cells for decades.

We didn't think of them as adult stem cells when we started doing bone marrow transplants, but that's what bone marrow transplants were when they were first done in the early 1960s. The reason those transplants took is that stem cells from the donor replenished the blood and immunological cells of the recipient.

So there is actually a lot of experience with the use in that setting, and there are a couple of other settings where I think adult stem cells could be quite useful.

One area is in cardiac tissue, for example, with heart attack victims, where if you could get adult stem cells to that location and you got them to behave properly and tractably, you could replenish some of the damaged tissue that occurs during a heart attack. There are likely to be some tests on this. There has already been some work done in Europe on this concept, and there is likely to be some in this country. I know one company was thinking of trying to get a study started as early as this year.

The reservation about going whole hog into adult stem cells stems comes from a couple of things. A couple of years ago there was a suggestion you could take blood stem cells, for example, and you could convert them into brain cells, or neurons. That work has looked a little bit more problematic on further revisitation over the past couple of years. So there is uncertainty about that, I think, would be the fair way to put it.

So the versatility of adult stem cells, I think, is still an open issue, and really needs to be addressed.

The interest in embryonic stem cells, and the reason there was so much debate about it, and so much to-do about it, is because these are blank slate cells which have the capacity to go in any direction. If you learn how to nudge them in the proper direction, you could address, literally, over two hundred different cell type scenarios, or dysfunctions, with embryonic stem cells—if that power is harnessed.

I don't think that same versatility is offered by adult stem cells. The danger in putting all our eggs, as it were, in the adult stem cell basket, would be the following: we work five or ten years, which is a legitimate timeframe, to do just adult stem cell work, and at the end of that ten years, when we haven't done the embryonic work, discover that these things didn't work the way we were hoping they would work in a certain clinical situation. Then what do you do but to go back to the people who have Parkinson's, Alzheimer's, and diabetes, and say, "Well, we thought adult cells were better, but we were wrong."

KONDRACKE: How much money is being invested in adult stem cells, as opposed to embryonic stem cells?

HALL: There is some money—you know, it's interesting. I think the big money is not being invested. The big money is essentially from pharmaceutical companies, which are, I think, basically sitting on the sideline. They are watching this very carefully.

But I don't think they feel that this is close enough, right now, to jump into with both feet, partly because the science hasn't been worked out. I think it's going to take a little bit longer than a lot of people suspect, especially with the embryonic stem cells.

The whole other issue is FDA or regulatory approval of these therapies as medicines. These will be living cells which, at least initially, have this protein power to change, and that's going to put enormous pressure on any kind of company or research to be able to prove that the cells they are putting in are exactly the cells they say they are—that they are not contaminated cultures that have cells that are in different stages of development that might be going off in different directions. I think that is a very significant hurdle, which it's probably premature to talk about a lot because we are not there yet. I think that's going to be a big issue.

KONDRACKE: Well, since President Bush limited the amount of research that can be done with embryonic stem cells, is the National Institutes of Health spending a lot of money on adult stem cell research, and if so, how much?

HALL: You know, I don't know the exact figure. I know they are spending a lot on it. They have tried to spend a lot on the embryonic stem cells as well.

But the issue, from the scientists that I have spoken with, is that with the limitations of the cell lines that are available, and we can talk about that a little bit more, it constricts the number of cell lines you can investigate. It's funny, I talked to Leonard Hayflick—one of the things I did is I talked to a number of the people involved in this story about what they were doing on the night of the president's announcement. Hayflick was very vocal in finding this policy inexplicable, and I am sort of softening his language. I invite everyone to read the book to get the actual, more graphic, language! But as someone who has been in cell biology for forty years, he knows especially well that each one of these cell lines is almost like a child in the sense that it has different characteristics, it has different traits—it is easier to grow, it's easier to raise, it's easier to do this or work in certain areas, it's not so good for other things. Ideally, you would like as many as possible, as soon as possible, in order to find the ones that seem to lend themselves best to the basic research that really needs to go on now. I think the fact that they weren't available has slowed things up.

I think the fact that the NIH, to a certain extent, was a party to this whole issue about the number of stem cells that were actually available, and then weren't. We can talk about that a little bit, as well.

I detected a certain degree of wariness on the part of people that I would consider to be very good stem cell researchers about the NIH's role in this, and whether politics were not influencing the way these things were being handled.

KONDRACKE: I want to get into that, but I am just trying to understand. If the Bush administration and the conservatives are serious about adult stem cell research, is the money following it? Is there private money in adult stem cell research so that whatever the potential of that line of research is can be fulfilled and we will have a good idea of how far that can go?

HALL: To answer the general question, yes, they have supported that research. It's interesting, though, in terms of the private money involved in the companies pursuing this. This is basically my feeling, and venture capitalists have told me the same thing, controversy chases money away. Even though adult stem cells are different from embryonic stem cells, they are stem cells. People know that stem cells are controversial and that they are always in the news, and people are arguing about it, and there is a constant debate about it.

I think that has scared some money away. I note this at the end of my book. You know, a lot of the development of both embryonic and adult stem cells, in terms of the private sector development, most of those companies are kind of struggling. They don't have a lot of cash on hand. They are not well capitalized. The money has stayed away.

The NIH has been funding the embryonic stem cell research for the cell lines that are existing, but again, researchers have actually been going and creating privatized situations. Stanford set up a stem cell institute. Some of the private foundations are funding researchers overseas.

There is a wariness about the NIH's intentions here, and I think that's affecting things a little bit, psychologically, if not literally the amount of dollars in play. I think people are a little bit wary and want to see where this is going to unfold.

One researcher told me, and this typifies the problem, he said: "I can't start a program in embryonic stem cell research now. I know it is going to take five years to unfold, and then three or four years down the line to find that politically the rug is pulled out from under me, and I won't get any post-graduates or graduate students to work on it, either."

KONDRACKE: Well, as everybody probably knows, and for the sake of those who don't, what President Bush announced on August 9, 2001, was that federal funds would go for research involving stem cell lines that had already been derived from "leftover embryos" in in vitro fertilization clinics; that is, the embryos had already been destroyed. The stem cells had been taken out from them, and the lines had been developed. He claimed at the time that there were sixty-four such lines—sixty lines, I believe he said, and then the government said, "It's actually sixty-four."

HALL: Right.

KONDRACKE: And people have been chasing around trying to find these sixty-four lines ever since, and there are in fact how many lines?

HALL: I think it's now up to twelve. The history of this is actually very interesting. I actually think it has not been talked about too much in terms of how it affected the relationship between the scientific community and the NIH in this area.

The president made the announcement and said that there were more than sixty cell lines available. From what I'm told, that language in the speech was never vetted by the NIH. They never had an opportunity to suggest that that may have overstated the case.

I was also told that the people at the NIH were flabbergasted when they heard the president mention that number, because their understanding that there were sixty-some cell lines, but they were in various states of development, and the state of development in a cell line is basically you can't call it a cell line unless it's established or not.

There was an immediate reaction from the scientific community because they knew there had only been about a half a dozen cell lines published in the literature that had been established as bona fide real stem cell lines.

There was, as you recall, I'm sure, a huge outcry about this, over how many cell lines there really were. And the press started tracking down every one of them.

What I found really interesting about this, and I actually did an op-ed piece for the *Times* a couple of months ago, is that the parallels to the weapons of mass destruction argument in the Iraq war are kind of haunting, because the people in the White House knew that not all these things were cell lines, from what I'm told. They had a very good scientific understanding of what was being discussed, but somehow the president was allowed to say that there were more than sixty lines.

Now, you could say that the exchange of information had been mistaken and so on. And they could have then said, "We misunderstood" or, "We really maybe perhaps overstated the case." But they didn't do that. Under heavy pressure and continuous pressure over the course of the next month, they didn't say, "We misunderstood it." They kept insisting, "The cell lines are robust. They are viable. They are ready to go." They continually made that argument for a month up, until there was a Senate hearing on September 5, which was about a month after the president's announcement, and under some very intense and relentless questioning by several senators, Tommy Thompson admitted that maybe there were only two or three dozen cell lines that were actually ready to go.

So only under the most forceful kind of skepticism did they even relent from the original argument. What was really interesting is that then a week later September 11 happened. From where I sat, I felt like this whole policy was beginning to unravel because there was a momentum—and you know this as a reporter, too—there is a certain momentum to stories when everybody starts jumping on it and all the stories tend to reinforce each other that there are not these cell lines. I felt like that was happening to the Bush policy in that it was really being severely challenged.

Then September 11 happened and, of course, it instantly dropped from everybody's radar screen, and I think it gave the Administration a buffer on this particular issue.

You have to remember too, this was the first major policy decision that the Bush Administration made. I mean, it was the first one that was the focus of really intense public debate, part of which was because of the protracted decision-making process that took so long. Everybody was talking about it and it just assumed it's own kind of weather, almost.

KONDRACKE: And then I think September 11 kind of interrupted that momentum. What happened after that?

HALL: Well, then the NIH finally listed all these available cell lines in November 2001. The total at that time was seventy-one, so it actually had grown larger.

Now, I was in touch with researchers who were trying to track these down, and these were good stem cell researchers, and they felt they needed to do their due diligence in trying to contact all the sources of these cells.

They spent days and weeks, if not months, contacting every single person on the NIH registry and they found that only one or two cell lines were available. The key word is "available," because it wasn't just that they existed, but that they would be available without strings attached in terms of the intellectual property.

Most of them never existed, and the notion that people are wasting their time on government grants tracking down cell lines that did not exist—I think that this where some of the wariness about the NIH grows into this. People at the NIH knew that these cell lines were not available, and they knew that they were not complete cell lines. And I think they knew that for well over a year.

KONDRACKE: So do you think that the president lied?

HALL: I think this is very complicated stuff, and I think it's easy to misspeak, and I probably have about fifteen times in the first half hour here myself. However, with that intense scrutiny, and with that intense challenge, I think that that information could have been corrected sooner. I think there were some political reasons.

I think the long-term political goal was to keep embryonic stem cell research going at a slow pace and hope that the adult stem cell story would go faster and that you could say, "Here, see, this will work. We don't need the embryonic cells."

KONDRACKE: Good. Now, we've got another level of controversy involving stem cells, which is cloning.

HALL: Right.

KONDRACKE: Therapeutic cloning. There is a bill that passed the House of Representatives that bans all cloning, both cloning of human beings and also cloning of embryos for research purposes. What do you understand the political situation in the Senate is? Will this bill ever pass? or do you think it's going to get stuck?

HALL: I actually thought it was going to come up in the spring after the House passed it in February, I think. Then it went to the Senate. The sense I was getting in the early spring was that it would come up maybe in late May or early June in the Senate. The fact that it didn't leaves me to suspect that the votes were not there to pass it, because they didn't want to have a discussion about something they weren't going to be able to get through. So what happens in the fall? It's not clear.

I will say this: as the House vote was driven in part by the announcement at the very end of last year by the Raelians that they had cloned a human being, these legislative events have been very much been driven by these public disclosures that kind of come out of the blue. That's why the privatization of this research, as it were, has been a real wild card in the whole public discussion of this.

To revisit that Raelian claim, there were an enormous amount of bioethical resources and journalistic resources and legislative resources expended on what has turned out to be, obviously, a non-event. I think in some instances the public debate on this has been whipped around needlessly by some of these claims that really haven't turned out to be true.

I would argue that the likelihood of being able to clone a human being for reproductive purposes, if not absolutely impossible, looks to be much more difficult than I think people suspect.

KONDRACKE: If you can do it with sheep, why can't you do this in humans?

HALL: The story in primates and probably in humans, if people have actually been trying to do it in humans, is that the process is slightly different. It's more difficult. The primate researchers are the ones who are saying because of several technical issues, this may not even be possible.

In fact, I think *Science* ran a piece by Gerry Schatten's group in Pittsburgh about this not too long ago. I know of some people who have been involved in mouse cloning have also raised the same questions.

Even in the case of animals, Dolly was one in 277 tries, and, in fact, may have suffered some developmental damage because of the process as well. So I am not even sure that you can do this. But, to say that you can't even do research on it because they don't want people to refine the techniques that might be applied to reproductive cloning, I think goes to an extreme.

KONDRACKE: Now, one of your arguments is that even though cloning is not illegal, and even though use of left over in vitro cells for embryonic research can go on in the private sector.

HALL: Right.

KONDRACKE: One of your arguments is that, in fact, the government restrictions have driven research out of the universities and more into the private sector, where there is less discipline, where there is less publication, more proprietary control over the fruits of the research and so on.

You contrast that with Britain, where there is something called the Human Fertilization and Embryology Authority. Now what does that do, and if we had one, how would it be better in the United States?

HALL: This actually grew out of the work in IBF in the late 1970s. The first test-tube baby actually just celebrated her twenty-fifth anniversary—or birthday.

The notion was, they wanted to keep track of these technologies. They wanted to monitor it. They wanted to know exactly how many embryos were being created. They wanted to know the success rate of the transfer procedures. And to this day they know exactly how many embryos have been created in England. They know where they are. They know what their status is. They know who's donated them for research purposes or whatever.

In other words, they have a very firm grasp on what has happened. Then with each incremental development of technology, there is an application to the authority, as in the case of Ian Wilmut, who cloned Dolly, who is interested now in getting into human therapeutic cloning. And that has to be applied for through that authority. So they basically keep track of, oversee, and regulate that whole area of research. It grew out of reproductive medicine, but it has extended into these new technologies.

At a similar moment in our sort of medical history—there was a great deal of discussion in Washington through the mid- and late-1970s about federal support for embryo research, fetal tissue research. A lot of people think this is a purely a Republican reaction against it, but actually it happened in the Carter administration first. What was then the HEW secretary refused to approve the notion of NIH funding this research and, for various logistical ledger domains that I talk about in the book, it never came up again until the early 1990s.

It was largely driven by this age-old right-to-life argument, which has inflected so much of the discussion, but was also driven by the medical research policy in this country, certainly in terms of reproductive medicine, and now into some of these regenerative medicine technologies, as well.

KONDRACKE: Right. You argue that all such research should go forward. Is that is your basic position? that at least embryonic stem cell research should be allowed to proceed? I take it you would like to have a regulatory regime, something like Britain's, to oversee it, as opposed to it just going willy-nilly as it might in the private sector?

HALL: You know, I think the NIH could probably handle that task. There certainly wouldn't be a lot of people in over their shoulder, given everything that has preceded it. I think the NIH could conceivably do that. I think there is sentiment for having a more formal regulatory structure for both reproductive medicine and stem cell research, and it is unclear how that is going to play out in the next couple of years. But there seems to be a movement in that direction.

I just think the NIH would be much better—not only as a kind of scientific arbiter, but there is also a lot of moral persuasion behind the NIH because they always get the best people in the field to oversee a field, and do the peer review, and that sort of thing. And they have clearly have been sensitized in this whole area. I think that way it would free up the federal money, and it also, I think, would actually make the research more rigorous.

KONDRACKE: Right. If I have one criticism of your book, it is that I don't think that you gave adequate attention to Leon Kass. You certainly mention him, and you say that he's profoundly reactionary, although you do say that he is also morally serious.

But the kind of objections that are raised to cloning, it seems to me, might have been answered more extensively than they were. I'll give you one specific that I don't think you mentioned at all, and that is the slippery slope argument.

HALL: Um-hmm.

KONDRACKE: In Britain they limit the growth of embryos to fourteen days, right?

HALL: Right.

KONDRACKE: So after that they can't be developed any more. Independent of the argument that we are going to clone babies and some crazy person is going to clone himself or something like that, what is to prevent someone from allowing a fertilized embryo to grow, not to fourteen days but to fourteen weeks, so that we can harvest its little heart for tissue or something like that? That is the kind of brave new world scenario that, if science is sort of left to go its own way unrestricted and so on, conceivably that could happen? That is something that people are afraid of.

HALL: Right.

KONDRACKE: What is the answer to that?

HALL: I think that's a legitimate fear, and I think that could be handled legislatively, if you want to put a fourteen day limit on development. That seems like a reasonable line to draw.

I think you can build in limitations and put limits on what you can do. I think you can attach penalties to it, for example, in the corporate setting, that would be essentially equivalent to a company-ending event, if they were to violate that limit. I think there are ways to handle it, short of having an overall ban on all the forms of research.

The slippery slope arguments I'm always wary of, and I do talk about this a little bit in the book. It is kind of timid and reactive to not do something because something else might happen.

I think you are much wiser to build a safeguard around what you know might happen, and then, if the circumstances change in a year or two because the research suggests different circumstances, then you can address it again. But to put a blanket ban on certain activities because they might lead to something may be being too cautious. If the Wright brothers thought that there might be a plane crash, and thought, "We better not develop this technology"—this is a gross and perhaps unfair example, but any time you are developing a new technology there are things that might go wrong.

I think the wiser course is to identify that which might go wrong and segregate it from that which is likely to benefit society, and use whatever tools are necessary, be they legislative of scientific community or public to shift it in the direction of benefit, and not in the direction that we don't want it to go.

KONDRACKE: So, at the present state of things, how do you think the United States is going to lose out on this research because of the restrictions that the government has imposed? and is it going to go off shore? Is it going to proceed at a less robust pace than it could in this country? What do you think the bottom line is here?

HALL: I think it is proceeding in a less robust pace than it could be in this country. I mentioned the example of the professor who said he didn't want to start a project that might be intercepted at some point.

One of the things I hear from some stem cell researchers, whose work I respect, is that they can't get young people to work in these areas because it's politically too risky.

I think probably your audience understands, but a lot of people in the public don't understand, that the vast amount of cutting-edge research is not done by the lab head whose name is on the end of the paper, but it's done by the people who are working in the labs eighteen and twenty hours a day, who are solving the incremental daily problems that need to be solved for any experiment to work.

Those are the people who do the bulk of the scientific innovation in this country. If they are scared away from this field because of the political uncertainty that is attached to it, we are really losing this incredible resource that we have, and have enjoyed for such a long period of time. They will go to other places, such as Sweden, Israel, China, Singapore, and England, where there are fewer restrictions on this avenue of research and people are going full throttle.

KONDRACKE: Right. OK. Let's do a little audience participation here. These are questions that have either been e-mailed in or are from the audience here: "With the private sector increasingly becoming the owners of genomic research, do you think that in the future scientific breakthroughs could be economically rationed to the highest bidder?"

HALL: Well, in a sense, they already are if you consider the drug industry and the price people pay for pharmaceuticals, and the prices that are set in this country, for example, versus the price for the same medicine in Canada or in Europe.

So we are the highest bidder. We already are the highest bidder. Part of it is built into our health care system and delivery of it. That will only continue, and believe me, there will be a premium on any of these medications if they are shown to be effective.

KONDRACKE: "Do you think that scientists are in tune with the real-world implications of their research—its effects on public policy, population, culture and human interaction?"

HALL: I do. I think scientists are much more aware of the public ramifications of their work than the public is aware of the political ramifications of some of the things that they suggest in the scientific community.

You can't function in many of these areas without having institutional review boards, and there is a lot of inter-community scientific interaction on all sorts of levels. I think people are very sensitive, particularly on issues where there has been a big public debate, as in stem cells. They are very sensitized to these issues.

I don't think the public realizes what a blanket ban on scientific research means, not only for the specifics of the scientific research, but it has some ominous intellectual and historical parallels to things that we haven't talked about in centuries, like Galileo and the Vatican church.

I mean, restricting human curiosity, although it will probably not figure high on anyone's list of daily priorities, is really a critical issue that underlies the whole structure and monument of American biomedical research for the last fifty years.

KONDRACKE: One of the things I find curious about this debate is that some of the people who are arguing against cloning, for example, the neoconservatives—Bill Crystal of the *Weekly Standard*, Charles Krauthammer, and I guess even Leon Kass is a bit a traditionalist—who believe that dying at the right time is a good thing. Basically.

HALL: Exactly.

KONDRACKE: But under ordinary circumstances, these are people, the neoconservatives, who believe in free inquiry, capitalism, all this kind of stuff. So what is it about their position? What was the intellectual background of their objections?

HALL: You know, it's an excellent question, because it actually is so contrary to everything else in the philosophy.

KONDRACKE: And they are not right-to-lifers. I mean, they are not particularly religious people.

HALL: Not particularly. But there is a social agenda, a family agenda, and a moral agenda, and there's some real interest in kind of telling people what's right and wrong. I think this informs a large part of their interest in this area. And I think technology does frighten people, and one can concoct frightening scenarios out of some of these technologies. There is no question about it.

I think the proper response is to not abandon the technology, but to control it in a socially acceptable and correct way.

KONDRACKE: OK. As a journalist, what do you see as the media's role in covering the issue of "immortality"? and how does the coverage drive public opinion on controversial topics such as stem cells?

HALL: You know, as I mentioned in the book, toward the end, the—when I started out talking with the title, *Merchants of Immortality,* I was thinking primarily of the privatization and the commercial interest in developing these technologies.

As I worked my way through it and saw what other people are doing, and what I was doing, and some of the public arguments, I realized that the merchandizing of the idea of immortality extends to those of us in the media. We are guilty of taking a kind of slender idea, a possibility, and kind of giving it more credence by mentioning it all the time. And I see it in work that I have done, so there is certainly that in play.

You mentioned Leon Kass. I think this whole notion of finitude, and complaining or arguing against immortality is using immortality to merchandise his ideology, which is not just for finitude and not immortality, but as I quote in the book, he has some anti-scientific sentiments that are really surprising in this day and age. They go beyond concern and wariness, but they actually suggest that science, because it does search for truth and searches for the correct answer to things, is a kind of socially destabilizing force.

KONDRACKE: We're just about done. I have one more question. What is the number one barrier to "immortality"? Politics, lack of knowledge, lack of funding?

HALL: I would say actually, the number one barrier might be natural. I don't think we are designed to live forever. I think we can extend life a little bit, but even doing that is going to take an awful lot of work in the laboratory.

KONDRACKE: Well, thank you. Steve Hall, thank you so much for being with us. This is the book, *Merchants of Immortality*. I highly recommend it and thank you all for being with us. Thank you.

END

The War on Antiaging Medicine

Robert H. Binstock, Case Western
S. Jay Olshansky, University of Illinois
Morton Kondracke, Moderator
September 17, 2003

For more information on debate participants and SAGE Crossroads go to
www.sagecrossroads.net

KONDRACKE: The shorthand subject of today's discussion is the war over antiaging research. We will get beyond that issue, but that's the headline of the discussion.

The discussants are Dr. Robert Binstock, who is professor of aging, health and society at Case Western Reserve University, and S. Jay Olshansky, who is a professor in the School of Public Health at the University of Illinois at Chicago, and a research associate at the Center on Aging at the University of Chicago and at the London School of Hygiene & Tropical Medicine.

So the opening question, which I would like each of you to respond to, is what is the war over antiaging medicine about? Who are the combatants? and what are the stakes? Jay Olshansky, you can begin.

OLSHANSKY: All right. Let me give you a little bit of background, actually.

A couple of years ago at the American Association for the Advancement of Science (AAAS) meetings, we were having a session on how long humans can live. At that session there were a number of scientists who were talking about duration of life and how much further we could extend it.

Many of us were lamenting, though, about the prospect of some individuals suggesting that we already have the ability to slow down, stop, or reverse the aging process. We were suggesting that that is not currently possible. Although that is what we know is not currently possible, it may be at some time in the future, but we know it's not now.

That was the beginning of something that my colleagues and I turned into what was called the position statement on human aging that we published in *Scientific American* last summer. It was also followed by a short piece in the June issue of *Scientific American* called "No Truth to the Fountain of Youth."

We were basically making the argument that there are a number of individuals out there who are currently selling products with the specific claim that it is now possible to slow, stop, or reverse human aging. We wanted to state in language that was as unambiguous as we could possibly make it, that not only is that not true, but there are some potential dangers associated with some of the products that are being sold.

At the same time, we wanted to emphasize that there is an ongoing and fascinating field of research in gerontology that is devoted to the study of aging, with the goal, in part, for some of the researchers, of understanding the process of aging and finding a way to intervene so that at some time in the future we might be able to slow down the process.

So we wanted to clearly distinguish between the substances that were being sold today with claims that it is currently possible to slow, stop, or reverse aging, with the real science of aging, suggesting that at some time in the future it might be possible, but it is not currently possible.

And the position statement on human aging lays out in very clear language what we know, what we don't know, what we think is true, what we think is not true, and we had quite a response to it.

KONDRACKE: I take it that one of the stakes involved here is that you are afraid that the reputation of gerontology will somehow be affected or clouded by antiaging quackery. Is that true?

OLSHANSKY: Well, there are two things. First of all, we wanted to get across a public health message. It is a public health concern that individuals are using products that have not yet been tested for efficacy—whether they do what these individuals say that they do—nor have they been tested for potential harm.

So first it was a public health message.

But as you said, we were also concerned with the possibility that the research scientists who were doing work in this area might be associated with this antiaging industry. And of course, the antiaging industry, as many have documented, has been with us for thousands of years. It appears to crop up right about the time there are major developments in our ability to influence diseases. Antiaging medicine, for example, was extraordinarily popular at the beginning of the twentieth century, as we made inroads against infectious and parasitic diseases.

It is no surprise that it's cropped up in today's world during a time in which we are aging very rapidly as a population, as we see major advances that are occurring in the biomedical sciences in the form of stem cell technology and so forth, that seem to be very promising.

So it's consistent. It's exactly what we would expect to occur at this particular time.

KONDRACKE: Dr. Binstock? What do you have to say about this?

BINSTOCK: Well, I got into this subject matter a couple of summers ago when Jay and his colleagues published this "No Truth to the Fountain of Youth" article, and I asked myself, why are they doing this?

After all, antiaging interventions have been around for millenniums, and there have been plenty of fraudulent, risky, and harmful products out there.

And when I did, it seemed to me that there were, indeed, these two strains. One of them was the public health message. But the other one was very much boundary work, which you alluded to with your question.

On the public health front, this is an important thing to address. The size of this industry is estimated by one report I've seen to be about $42 billion, if the market is broadly defined to include exercise and nutritionals, as well as cosmetics and hormone injections and dietary supplements. That estimate also predicts that it will reach about $67 billion in a few years.

And there is no question that some of the practices are risky, harmful, and fraudulent. For example, one of the things they push very much is hormone replacement therapy. Well, you probably have seen a number of the studies recently pointing out not only the physical, but mental risks of that therapy.

So I saw this as a worthwhile undertaking. But I was also cognizant of the fact that research into biology of aging has a long history of being regarded as being a marginal, charlatanic undertaking itself. And here were over fifty international biogerontologists coming out and making this statement: "No Truth to the Fountain of Youth."

So I went back and interpreted and documented the history of how aging research was regarded very poorly by the NIH, and by all sorts in the federal government, even as people in the field tried to establish their own National Institute on Aging in the early 1970s.

The statements from NIH officials, from the Office of Management and Budget, and from assistant secretaries and so on were just as damning as you could be:

"These people are not competent. They don't have any good ideas. No good will come from it. It's a pipe dream," and so on.

Now some of this opposition, of course, especially from the NIH, was that they didn't want another institute sharing the pot of NIH appropriations. But in fact, this was very much the view.

Nonetheless, the biogerontologists, in particular, persisted with their lobbying to the point where they managed to get a bill passed in 1972 to create their own trough for their research funds, which was why they said that everyone was prejudiced against them.

And President Nixon vetoed it, then, on the advice of NIH and OMB. But then they persisted and got it in 1974, and of course, President Nixon was wrapped up with other things, such as Watergate and his impeachment.

So the bill was left to pass, and that began a legitimation of this field, which has been extraordinary since the institute began operating in 1976. It's grown from what was then about $50 million to a billion dollars today.

So it was very important for the biogerontologists not to be tarred with the brush of this antiaging medicine movement. Indeed, when I wrote this up, I got quotes from some of the leaders, some of Jay's colleagues, who said, "Indeed, this was an important purpose."

I went and examined then, how is this war going, this attack on the American Academy of Antiaging Medicine (A4M), in particular? It claims twelve thousand members. It board certifies people as antiaging practitioners, even though it doesn't have any approval from the American Medical Association. Its net assets have grown from $500,000 in 1997 to over $5 million today in terms of their IRS report. So it's a thriving, growing thing and the question then becomes, what's the best strategy for trying to combat them? They are risky and harmful and fraudulent practices. Well, Jay has undertaken one strategy with his colleagues.

KONDRACKE: A frontal attack, you would say?

BINSTOCK: To the fountain of youth, right?

KONDRACKE: Right.

BINSTOCK: He has also issued a couple of annual silver fleece awards emulating Senator Proxmire's golden fleece awards, which he gave out years ago.

And as I assess the war on antiaging medicine, as they began trading shots, one of the things that struck me was that this strategy may not be the best, to attack this movement, because a Brer Rabbit type of situation has emerged. They are getting inextricably involved with a tar baby.

For example, recently the *O'Reilly Factor* show wanted to deal with this subject, and the head of the antiaging medicine movement, as misleading as he is in my judgment, was given the same platform as Jay. A month after the "No Truth to the Fountain of Youth," came out in *Scientific American*, the *AARP Bulletin* ran a story on its front page about "No Truth to the Fountain of Youth," and in it they have Ron Klatz, the president of A4M, on an equal platform with Jay. There was a nice big picture, a sidebar, and indeed, the very fact that the reporter felt it was important and imperative to go interview Klatz and get his view, told me that this group is getting legitimated a great deal, at least, in the eyes of the popular press.

So I won't go on at length about this, but I think there are maybe better strategies for the biogerontologists to pursue.

KONDRACKE: Such as?

BINSTOCK: Well, first of all, I think that to avoid conflating what biogerontologists are working on, which is the fundamental causes of aging and how they might be used constructively as we understand them, they would be better off emphasizing that what they are up to is trying to achieve active and healthy long life and hit on that rather than saying "These other guys are bad to accentuate the positive."

Now, the question of what to do about the risky and harmful and fraudulent aspects of antiaging medicine is very tricky, because stronger government regulation doesn't seem to be in the cards. One of the things that has really led to the growth of this industry was the passage of legislation in 1994 that protected dietary supplements from any sort of government regulation.

The FDA and the FTC can't regulate the practice of medicine. To the extent that that's done, it's done by the states. The states have shown no sign of getting into this issue so far. And the power of the dietary supplement industry in being able to get that legislation in, and keep it in force for ten years, suggests it is going to be pretty hard to overturn.

So my feeling is that the ball is really in the court of the professional organizations that care most, or purport most to care, about the well being of older persons—namely, the Gerontological Society of America, and the American Geriatrics Society—that as organizations they should address antiaging medicine, perhaps, through specific task forces that could work jointly and singly, and I can think of at least three things that they should do.

One thing they should do is try to embrace many antiaging practitioners by having symposia and workshops at their annual meetings, and reaching out to them and helping to sort out the good and the bad and the ineffective for them, because I don't believe that everybody out there practicing in this area is doing bad things consciously.

KONDRACKE: Uh-huh.

BINSTOCK: Or being quacks consciously.

A second thing they can do is identify those areas of antiaging interventions which are problematic, and designate their experts to do reviews of them in health care journals, such as was done this year on human growth hormone in the *New England Journal of Medicine*, reviewing its harms, its assets and what the status is now. And in a sense, be police on this by getting these reviews done in a proactive way.

KONDRACKE: So this role would not be taken by the biogerontologists themselves, the academic biogerontologists, but organizations that represent them?

BINSTOCK: Well, in this case I would say the professional organizations that include the biogerontologists and the geriatricians and so on.

But my third step is that they should reach out to the others. For example, these professional organizations could develop white papers and arrange for press conferences at the National Press Club, orchestrating in the AARP, Consumers Union, the American Medical Association, the American Public Health Association, so that the word about the harms and the risks and the fraudulency could get out through the popular media much better than its has to date.

KONDRACKE: Let's go back to Jay to give him a chance to respond to your proposal here.

OLSHANSKY: OK. Well, actually, you know, we were talking about this earlier this afternoon, and it's interesting. I agree with Bob. He raises a really important issue that I think we were unaware of when we were writing this position statement. And that is, once it got out, and once it became publicized, there were stories in newspapers and magazines that were giving equal time to the very people we were saying really are not scientists in the field and don't deserve to be given equal time. And that is a problem. I agree with that. It was an unintended consequence of what we had done.

KONDRACKE: So you are not going to go on the *O'Reilly Factor* any more with Ronald Klatz, the head of the American Academy of Antiaging Medicine?

OLSHANSKY: Actually, it's hard for me—you know, if given an opportunity to debate, to turn it down. I am not afraid to confront the opposing point of view, and I certainly am not afraid to oppose the other point of view.

So I will always accept an opportunity to debate under these conditions. Someone has to do it. Someone has to stand up and state specifically what we know and what we don't know in light of what they are arguing.

So I agree on the one hand. Now, the other thing that Bob was recommending was the creation of these white papers and interventions by government agencies, and so forth, to make it clear what we know and what we don't know about aging. In a way, that is exactly what we were doing with our position statement, except we were focusing in on the scientists themselves who are doing the research.

If there is going to be some sort of breakthrough in the field of regenerative medicine or antiaging research, it is going to come from many of the scientists who had already signed on to our position statement.

So my view is that I don't think it is going to work. I think what would happen is if we write these white papers, we would get the exact same effect that we got with our position statement, and that is that these alternative points of view about antiaging medicine are going to come to light, more people are going to become aware of them, and I think the antiaging industry will continue to grow.

My approach would be a bit more draconian, I guess. That is the approach that I think is currently being taken with regard to the downloading of music and Napster; that would be to sue, to go after these individuals, these groups, where specific claims are being made about the antiaging properties of these products. They have to demonstrate efficacy. They have to demonstrate that there is no danger. If it cannot be demonstrated, then this kind of information needs to be made available to the public. I think fear will work very effectively for this antiaging industry in protecting the general public from many of these products.

KONDRACKE: Now, we have two definite cases where products for which great claims were being made: ephedra and fen-phen. But therapies such as hormone replacement have had major attacks made on them scientifically.

Now, one would think that stimulating science on these various products, one by one, or the most dangerous at a time, would somehow be a way to warn the public that there are problems here.

OLSHANSKY: Absolutely. You are recommending what I think would be the best approach, and that is to evaluate these products one at a time. Fen-phen and ephedra are classic examples of products that were sold to the public, some of which where sold by clinicians, suggesting that, "This is a reasonable product to be using. There is no danger associated with it." The fact is that it had yet to be evaluated using clinical trials in humans.

These substances that are being sold, the antioxidants, the growth hormone, the DHEA, all of these need to be evaluated in detail using clinical trials in humans before they are used, before they are administered.

My personal opinion is that clinicians, physicians who are administering growth hormone, a biologically active compound to their patients, before this product is adequately evaluated using standard scientific methods is inappropriate.

KONDRACKE: Now, the law, as stated by Dr. Binstock, is that none of these therapies, including growth hormones, can be regulated by law at the present time in any way. Is that correct?

What should be done? What is the ideal circumstance? Should the FDA be given responsibility over all of this? There is an Office for Alternative Medicine at NIH. Should it be given more money to evaluate these supplements? How would you design this if you could start from scratch? What would you tell Congress to do?

OLSHANSKY: Well, I think there is no question that some of the initial research that is being done, that has been funded by NIH, on this very issue should be expanded. There was a study that came out last year by Blackman. It was published in a major journal on growth hormone. This was one of the first and best clinical trials that was done on growth hormone in humans.

But it is an initial study. They used a larger dose of growth hormone. These kinds of studies need to be repeated using lower doses. They need to expand the study population to evaluate them. But you have to realize that hormone replacement therapy in women has been evaluated for decades, and the science keeps waffling back and forth. You know, "Should we use it? Should we not use it?"

Growth hormone has not really even yet been studied. So it's a classic example of one of those circumstances where individuals associated with A4M are suggesting that, "We can't wait for the science. If we wait for the science, we'll all be dead. So we should be taking these products now and hope that the science will eventually catch up with us."

That's the line of reasoning that's being used.

BINSTOCK: See, where I vary on this is not in objecting to that in principle, but the question is, who's going to see that it gets done in a blanket enough way to be effective?

That's why I urge that professional societies, particularly those concerned with older people and aging baby boomers and their welfare, should take it on as their responsibility to see to it that their members start working on this, to get them interested in these sorts of clinical trials and so on.

I mean "should" is one thing, but then who is going to implement the "should"?

KONDRACKE: Right.

BINSTOCK: With respect to the lawsuits, just one comment. It's not clear to me how that's going to work. Maybe it can. Your only route is tort suits and proving the damage might be very, very difficult.

KONDRACKE: Well, if you are a baseball player or a football player and you take ephedra and you die, can't your surviving relatives sue the manufacturers?

BINSTOCK: Sure. But I am not sure yet we have cases of people dying from these substances.

KONDRACKE: Oh, yes, we do. Well, claim on ephedra that a football player died, I believe, or a baseball player. I can't remember which one it was; I think it was this summer.

BINSTOCK: Oh, that's right.

KONDRACKE: Right. OK. Ronald Klatz, who is the president of the American Academy of Antiaging Medicine, is not with us today so I want to just take some of the claims that are being made by his organization, which is the advocacy group that we are talking about and ask you about them.

They say the official definition of antiaging medicine is that it is a medical specialty founded on the application of advanced scientific and medical technologies for the early detection, prevention, treatment, and reversal of age-related diseases.

Now, scientific. To what extent are their activities scientific? That is, peer reviewed, scholarly, etc.?

OLSHANSKY: Well, clearly that particular comment, and many of the other related comments that come from A4M, represent classic examples of preventive medicine–identifying diseases, trying to prevent them, postpone them, delay them, treat them.

I think what the American Academy of Antiaging Medicine has done, and in fact what the antiaging industry has done, and the mistake that they have made, is that they associate aging with disease. This is a fundamental problem.

They believe that if you increase muscle mass, reduce the rate of bone loss, improve skin elasticity, and reduce your risk of heart disease, cancer, and stroke, that you are altering aging.

The fact of the matter is that you are treating the manifestations of aging. Even if we could somehow come up with some hypothetical magical cure for heart disease or cancer or stroke, those three diseases would be replaced by three other major causes of death, and aging itself would remain uninfluenced.

So they are basically narrowly defining aging as what they do in clinical science to treat disease. What we are suggesting is that that is a fundamental misunderstanding of efforts, to go after the biology of aging itself.

KONDRACKE: Well, no, but aren't they saying that if you take vitamin E, that you are not going to cure heart disease, but what you are going to do is forestall it, right? They say that the most effective thing you can do to postpone aging is to take multivitamins every day.

OLSHANSKY: Well, I think much of what they are suggesting is—I hate to put words in their mouth—but that some of the nutritional supplements that are being offered indeed will reduce the risk of some diseases. I think there have been trials that have demonstrated that some diseases are amenable to modification with the introduction of nutritional supplements.

And that's certainly expected. There are plenty of things that you can do to influence attributes associated with aging and the risk of disease.

For example, many of the benefits that have been associated with the use of growth hormone, such as increased muscle mass, improved mental acuity, improved skin elasticity, all of these can be accomplished with exercise. Does that mean that exercise is slowing or reversing the aging process? No. It means that the manifestations of the aging process are inherently modifiable because the aging process itself is not programmed. In the absence of a program, it means that interventions work. That's precisely why many of the things that we do at older ages to treat the manifestations of aging enable us to live longer, and in some cases healthier, lives.

But that's not the same as going after the aging process itself. I think that is a fundamental misunderstanding of the antiaging industry.

BINSTOCK: I think to a layman the distinction that Jay is making is perfectly valid. But I think to a layman it is not terribly important.

As I said to Jay in a recent conversation, whether you are dealing with aging itself as a process or not, is not terribly important to a consumer like me. So when I do exercise, weightlifting and other things, to try to compensate for the condition that normally accompanies aging, which is loss of muscle mass, that could perfectly well be called an antiaging effort on my part. If my doctor prescribes it, it is a perfectly good thing to do. To me, the fact that it doesn't have anything to do with the fundamental biochemical mechanisms of aging is rather irrelevant.

So I think the key thing is to not focus on the industry as a whole and say it's making bad claims across the board and doing bad things across the board; it's important to sort out the good, the bad and the fraudulent. And there is plenty of good. I think there are a lot of practitioners out there who are advising exercise, nutritional stuff, dietary supplements as well as good diets, and they ought to be embraced and that tilting at them is not a good thing to do.

KONDRACKE: One of their other claims is that, indeed, since its creation in 1974, the U.S. National Institute on Aging has spent more than $9.4 billion, but has yet to turn any medical intervention into a meaningful application to combat the degenerative diseases of aging. So they are basically saying that scientific gerontology has produced nothing in all this time.

OLSHANSKY: Well, think about that statement for a second. I mean, it contradicts the opening remark which says that there are plenty of things that we can do, and that they do to intervene in the aging process, most of which was discovered by the very scientists they are saying haven't produced anything. So it is a contradictory statement and makes no sense to me whatsoever.

KONDRACKE: Does A4M lobby for or against increases in the National Institute on Aging budget?

BINSTOCK: I don't think they do either. But they attack the NIA very specifically as trying to maintain an establishment and to shut them out. They don't apply for any research funds that I am aware of.

They describe gerontology as a death cult. So basically, what they are saying is, "Anybody who picks on us is just trying to maintain power." In fact, that's what they described you as doing, to stay on top of the multibillion gerontological industry, and that's why you were attacking them, and so on. So they are good symbol manipulators.

KONDRACKE: Do they do any science? That is to say, do they have double blind studies? Do they publish peer-reviewed research? Do they test anything?

BINSTOCK: None that I know of.

OLSHANSKY: The leaders of the organization haven't published any scientific articles in a peer review journal. They have published plenty of popular books that all basically say the same thing—that it is possible to slow, stop, or reverse aging.

They will say that many of the people who attend their meetings, or who are affiliated with the organization, have published scientific articles. But the society itself doesn't publish anything that is in peer review.

KONDRACKE: They claim as sponsors, such organizations as Genentech, Novartis, SmithKline Beecham, and Tufts University. That sounds like legitimization. Is that valid or invalid?

OLSHANSKY: Well, I don't know if that's valid or invalid. I also don't know about any of those affiliations, whether those are real or not.

I think what you often see at their meetings is they will invite in well-known, established scientists. They will pay for their travel expenses and then they will give a keynote speech at their meetings. Then they claim association with real scientists who are doing real research on aging. And that, I think, is how they developed some sort of legitimacy—by linking up with scientists who are doing real scientific research.

BINSTOCK: My look into it indicates that at first blush they are able to get a little bit of money out of these various parties because, as you stated their goal, who could object to that goal?

KONDRACKE: Right.

BINSTOCK: That would be the goal of many geriatricians, by gerontologists and so on. So for example, the Retirement Research Foundation, which I believed funds this particular program in part, gave them an initial grant when they started up and left the sports medicine business in about 1993 for organizing a board of directors. They gave them $15,000, and so then they were able to say, "supported by the Retirement Research Foundation." They were just starting up saying, "We have these great objectives."

KONDRACKE: They list as promising areas of research—they don't claim that they are doing it, but they are advocating it—some of the same things that I guess gerontologists are looking toward as future possibilities for breakthroughs:

genetic engineering, including work with stem cells, cloning, nanotechnology, artificial organs, nerve impulse continuity. Are they trying to lay claim to some of the same activities that legitimate biogerontologists are undertaking in difficult laboratory research?

OLSHANSKY: Well, the vast majority of that list represents technologies that do not currently exist, and which do not currently influence the duration of lives of most people.

In fact, I think there have been only a handful of people who have been influenced by any of those technologies, and it may be years, decades, perhaps never, that some of those technologies will have some influence on aging.

Remember, you listed all of these potential technologies. None of them currently exist that have had any dramatic effect. And yet they are claiming that it is already possible at antiaging clinics that exist today, using the products that they are selling, to slow, stop, or reverse aging.

So you get this complex mixture of statements that it is already possible, and then a set of comments suggesting that the future will permit us to influence aging.

And so it makes perfect sense to me that they would do that, but clearly there is nothing that is on the market today that has, in fact, been demonstrated to influence aging in humans.

BINSTOCK: What's interesting to me about that list is that it doesn't resemble all the promising lines of biological research in gerontology, which in my view, are likely to make great progress in extending the human life span in a healthy fashion, and not in the far distant future.

KONDRACKE: Which are?

BINSTOCK: Things related to the successful experiments in caloric or dietary caloric restriction in animal models. There are scientists working on ways to mimic the biochemical effects of that so that people won't have to dietary-restrict themselves, and yet at the same time enjoy the same benefits that the rats and the mice have, which in some cases, as I think you know, Morton, is forty percent increase in average life expectancy and life span, from your interview with Rich Miller.

KONDRACKE: Um-hmm.

BINSTOCK: So that would be one example. There is the genetic manipulation work in the round worms, which looks promising. And there are a whole bunch of other things. But these things don't bear any resemblance to what biogerontologists are working on, or where the research really is.

KONDRACKE: Well, certain biogerontologists are working on stem cells.

BINSTOCK: That's right. That was one exception on the list.

KONDRACKE: Is this fight fundamentally about the efficacy of nutritional supplements and other nostrums on their part and what to do about them? Or is it a deeper philosophical claim about what is actually possible in aging research?

OLSHANSKY: I think most biogerontologists believe that there will come a time in the future when we understand enough about the aging process to develop an intervention that will slow it down, and most of us are optimistic that this will eventually happen.

In fact, to be honest with you, there is actually very little research going on focused specifically on identifying and going after the biological process of aging itself. The vast majority of the funds that are being spent are focused on specific disease processes, not the biological process of aging itself.

KONDRACKE: What percentage of the NIA budget is devoted to aging research—that is, the processes of aging—and what percentage toward diseases, such as Alzheimer's?

OLSHANSKY: Do you know the answer to this one?

BINSTOCK: Not off hand. I would take a guess that maybe the basic stuff is about twenty to twenty-five percent.

OLSHANSKY: Yes.

BINSTOCK: I'm not sure.

OLSHANSKY: I'm not sure. I will tell you, there is going to be a session at the forthcoming Gerontological Society of America meeting focused specifically on this issue, so I presume someone will provide an estimate at that time. I don't know what the percentage is, but it's far smaller than what is currently being spent on all the major fatal diseases.

KONDRACKE: Just on the philosophical level there are three—I believe one of you has written that there are basically three attitudes toward aging. One is compressed morbidity, which I will leave you to define. Another is decelerated aging, and a third is arrested aging, which is what the antiaging group is all about.

Now, would you define what these three are and what makes them different, and what the potential is for compressed morbidity, for example?

How long do you think we could extend the life span? What is it possible to do scientifically?

BINSTOCK: Well, I wrote about it, but Jay seems eager to answer it.

OLSHANSKY: Well, no, actually, why don't you talk about compression of morbidity.

BINSTOCK: The basic concept of compression of morbidity takes the fact that at the moment there is a maximum life span and says the goal—

KONDRACKE: Which is?

BINSTOCK: Well, it's estimated to be about 120 years, give or take things. The oldest woman in France died at 122.

Let's say the goal is to eliminate diseases and disabilities so that morbidity, the name for those, will be eliminated and people can live active, healthy lives until they reach the normal human maximum life span, and then just sort of fall apart and die like the old one-horse shay in Oliver Wendell Holmes' poem.

KONDRACKE: So would this push us beyond 120, or not?

BINSTOCK: No. It would push us up to where the maximum life span is without extending it.

KONDRACKE: OK.

BINSTOCK: It would certainly increase average life expectancy, however, because people wouldn't die of all these various diseases.

Decelerated aging is slowing the rate at which various aspects of aging take place, and they do it in concert in all the different species. This is the graying of the hair, the loss of the muscle mass, the loss of lung capacity, and so on and so forth.

So that's been accomplished in the animal model, such as in rats and mice. It has basically kept them healthy until near the end of their lives, but there is still a period of morbidity. What it has done is extend average life expectancy by roughly forty percent, and maximum life expectancy by roughly forty percent.

So Richard Miller, for example, says that through slowing the rate or decelerating aging, you would have the average life expectancy of 112 for an American Caucasian woman with an outlier of 140 years. His estimate.

Arrested aging is something that has been pushed very much by a geneticist at the University of Cambridge, although Americans have joined with him in this. The idea is to engineer negligible senescence, and maybe you can explain this better biologically, but to a layman, as I read the paper, what they are saying is, "We believe it will be possible to reach in and deal with the various biochemical things that cause these aging-related changes to take place, and reverse them before they cause damage." And that will arrest aging right there.

KONDRACKE: Give us an evaluation and tell me which side you are on.

OLSHANSKY: Well, this—

BINSTOCK: Do I pass?

OLSHANSKY: You get a C plus! And I'll talk to my students about this next week.

The issue of slowing and arresting the biological rate of aging is a difficult one to deal with. Let me just tell you; it is not currently possible to measure the rate at which aging occurs in any species. There is no measure. There is no suite of biomarkers, there are no magical whatever that we can use that will tell us how old we are biologically, and when we are going to die.

In the absence of measures of the biological rate of aging, it is therefore not possible for anyone who makes the claim that we have slowed aging, to defend it.

So what is used in its place, what is used in place of some sort of objective biological measure of aging, is something that is known as the actuarial rate of aging. It is known that the risk of death increases exponentially in humans about every seven years after the age of thirteen. You see comparable exponential increases in mortality among many other sexually reproducing species.

So aging has been defined by this rate of increase in the risk of death. When scientists have, through caloric restriction or genetic modification of one form or another, altered the death rates of these populations—and principally the way in which it is done is that the point of inflection, or the time period in the life span when the mortality rate increases, has been postponed from, but the rate of increase in the death rate for the most part remains the same—they have concluded that aging has been slowed.

What I would suggest, and this is actually the topic of a manuscript that my good friend and colleague Bruce Carnes from the University of Oklahoma and I are working on, is that we need to be questioning whether that is a viable measure of aging, the actuarial rate of aging itself.

So when someone says, as you have just said a moment ago, that researchers have demonstrated that we have slowed aging, I would throw a word of caution in there.

They have extended the duration of life of a wide variety of species. And we can extend the duration of our own lives by altering our lifestyles, by reducing our risk of a number of diseases, by wearing our seatbelts. Have we altered the basic biological process of aging? There is no evidence to support that claim at this time.

BINSTOCK: Well, you gave the demographer's answer. I give the layman's answer from reading the articles, which are entitled "Biomarkers of Aging," and "Reporting on What People are Finding."

So it seems to me that in the studies they have measured all sorts of things.

OLSHANSKY: Remember, there have been a number of studies on biomarkers of aging and the general conclusion is that there are no definitive biomarkers that enable us to measure aging. There has been plenty of research focused on biomarkers.

KONDRACKE: But you are a compressed morbidity person, are you? That is to say that that is what you think is presently possible? Or you think that decelerated aging is something that probably can be done but can't be measured at the moment?

OLSHANSKY: Well, if you are asking me what I would like to see happen, you know, this compression of morbidity argument was originally developed by Jim Fries in 1980. And, of course who wouldn't want to live very healthy right up until the moment in which they died? It is a desirable goal. Clearly it is something we would like to see happen.

But decelerated aging is something that, in my view, is—and you may find this surprising coming from me who is railing against this antiaging industry—but what I would argue is that instead of focusing so much attention on the manifestations of aging, the heart disease, the cancer, and stroke, and while it is, of course, important that we continue to focus on those diseases, we would achieve far more in the way of improved public health if we were to succeed in finding a way to slow down or decelerate the rate of aging.

Arresting the rate of aging, which to me implies that we are not aging at all, or that the risk of death is actually declining as we grow older, to me—

KONDRACKE: So say we find out biologically what causes aging, and learn how to do something about it.

OLSHANSKY: Learn how to do something about it to slow down the process. And the result would be, I hope, a simultaneous postponement of all of the manifestations of aging. So essentially, we would be younger longer.

KONDRACKE: Well, Dr. Binstock, I know, wants to say something about the societal implications of this, so let me ask a question about that.

Suppose we did any or all of that. We have extended life expectancy. We've got a lot more eighty-year-olds and one hundred-year-old people. And you know, there's all kinds of terrible demographic and economic possibilities from this, if we've got an aged population that is not working, that fewer and fewer young people have to support, etc., etc.

What is your view of the future?

BINSTOCK: I certainly think the implications of achieving this would be radical in terms of changing every institution we know.

However, I don't make the same assumptions that you do, such as that older people wouldn't be working.

If we have healthy, active older people, why wouldn't they be working? They'd want to be consuming things. We'd have all sorts of jobs and they would need to support themselves.

How could we support Social Security? Well, who says Social Security would be the same when that happens? In fact, there are some who are trying to make drastic changes in Social Security right now.

Would they all be politically together? These prolonged old people and the old, old and the young old and so on. Well, programs make constituencies, and so the way we shape programs might have a lot to say about that. Plus, they will have a lot of other identities than that. So extrapolation from what we know is about the worst mode of prediction, particularly in social affairs, particularly with respect to public policy.

KONDRACKE: It's a journalistic bad habit, I should tell you!

BINSTOCK: Well, no. Everybody does it and it's a great line for me to come out with.

But you know, this is very seriously happening. And I'd just like to make three points in very brief order.

The NIH has actively launched a program of trying to apply what we have learned from the caloric restriction experiment for applications to humans. They've got some pilot studies going on that with dietary restriction right now.

The implications of achieving anything like what's been done with animals and with caloric restriction or some of the other things are far more profound than the implication of cloning or stem cells. We are not even beginning to discuss it. I think if we begin to discuss this and debate its desirability or undesirability, as Leon Kass is posing, and what we can do to shape the future now through anticipatory deliberation, it would serve us all very well.

KONDRACKE: I think what we should do is have another session and bring you back on the societal implications. We have done some of that, but we'll do it in the future.

Let me just take a couple of questions that have been e-mailed in: "Many antiaging compounds, DHEA, human growth hormones, melatonin, etc., are natural compounds that can't be patented and thus are unlikely to attract funding from large pharmaceutical companies. Who will pay for the clinical trials to determine their efficacy?"

OLSHANSKY: That is a good question. I don't know. Actually, interestingly enough, I've had a discussion about this very issue with Ron Klatz from A4M, where I was suggesting, you know, if you know clinicians who are administering growth hormone and other hormones to their patients, why don't you enlist them in a trial? And he would say, "Well, we can't get funding for this kind of work." I don't know how hard they have tried to get funding, but—I don't know the answer to that question, exactly how that funding would be dealt with.

KONDRACKE: OK. Is it possible that the highly marketed concept of antiaging medicine could improve the visibility of gerontology? In other words, you have raised their visibility. Is it possible that they are raising yours?

OLSHANSKY: Not in a positive way. I think it's been in a very negative way, and I would clearly distinguish between this antiaging industry, which I refer to as an industry rather than medicine, and regenerative medicine, which is that whole list of technologies that may come in the future that will enable us to perhaps go after diseases much more effectively.

Then, hopefully, at some time many of the research scientists who I hope are listening, will focus in on the aging process itself and find a way to go after that. I would emphasize in terms of public policy that the age wave has yet to hit. It's not going to hit until the year 2011.

So as much as we like to consider ourselves an aged society, believe me, we haven't seen anything yet until after the year 2011. And for twenty years or thirty years after that, we are then going to see an aged society.

So I think if we don't attempt to focus in on the aging process itself, we are making a mistake, and there is as potential risk of dramatic increases in frailty and disability among the extreme elderly in the future.

BINSTOCK: I think that nothing but bad could happen from the antiaging medicine movement for gerontology because of conflating what Jay was just talking about.

Indeed, when the "No Truth to the Fountain of Youth" thing made the *AARP Bulletin*, the AARP message board was full of complaints from people saying, "There is truth to the fountain of youth."

So I think staying away from engaging with them and attacking them is the best strategy. I think there's a big tar baby there. The best thing is to promote one's self.

KONDRACKE: OK. We've got only four minutes left. So, does either of you have something burning that you want to say about any or all of this? Or else I will ask you a bottom line question, which I have asked before, but I just wanted to repeat it again.

OLSHANSKY: I'll take a question.

KONDRACKE: OK.

OLSHANSKY: Certainly.

KONDRACKE: This legislation was passed when?

OLSHANSKY: 1994.

KONDRACKE: 1994. Let's suppose that Congress revisits the subject. Let's propose that. What would that legislation say and do to correct the problem?

BINSTOCK: Well, I think the basic point is that you would try to get clinical trials done on all sorts of things that are not subject to them now, before they could be marketed.

OLSHANSKY: Bob and I were talking about this earlier. The question is, do you want to restrict access to these kinds of products by people who really want to use them? I think there is a real question as to whether or not such restrictions should be enacted. I'm not sure I believe that they should be enacted.

But the clinical trials need to get underway. We need to evaluate efficacy and potential dangers, and people need to be made aware of the potential dangers associated with these products.

I'm very concerned about clinicians, physicians at antiaging clinics making claims that are not substantiated by the scientific literature.

KONDRACKE: So would you create or empower an agency that already exists to begin testing and test according to what people's guesses are about what's the most dangerous, and start publishing results? Would that be an effective beginning at least?

OLSHANSKY: Yes. What is the organization that is already doing work?

KONDRACKE: There is an office at the NIH of Alternative Medicine.

OLSHANSKY: Yes, I think that would be a good place to start. Clearly, funding research through that organization may be on one of the best approaches.

KONDRACKE: OK. Thank you very much for being with us.

End.

Biomarkers of Aging: Do They Hold the Key in the Search for the Fountain of Youth?

David Harrison, The Jackson Lab
Dr. Roderick Bronson, Harvard Medical School
Morton Kondracke, Moderator
October 28, 2003

Pictured: Dr. Roderick Bronson, Harvard Medical School; Morton Kondracke, moderator; and David Harrison, The Jackson Lab.

For more information on debate participants and SAGE Crossroads go to
www.sagecrossroads.net

KONDRACE: Thanks very much. Our guests are David Harrison, who is a senior staff scientist at the Jackson Laboratory at Bar Harbor, Maine, and Rod Bronson, who is a pathologist in the Rodent HistoPathology Core at the Dana Farber Cancer Center at Harvard.

The basic question today is whether there is a way of tracking aging. That is to say, whether we can find a marker that will predict how long an organism is going to stay alive. This leads to the question, are there one or two processes of aging that we can put our finger on and watch over time? and will this have some predictive capacity?

That's the subject of today's debate, and it gets into the question of whether one is optimistic or pessimistic, and whether we'll ever get to the bottom of what aging really is.

So I'm going to let David Harrison, who is the optimist, start off with five minutes, and then Rod Bronson will answer back. Then we'll have an exchange. We welcome audience participation. Just write your question down on a piece of paper or e-mail it in, and we will be glad to put it to the guests.

Go ahead.

HARRISON: OK. I'm Dave Harrison from Bar Harbor. Thank you, Mort.

The purpose of the discussion here is to focus on biomarkers. Now, the information I was sent suggests that there are people who feel the NIA initiative on biomarkers was a failure. I find that hard to believe because the fact is that they did a lot of good, solid, basic research through that initiative.

The fact is, we do not have biomarkers that accurately predict life span in any creatures, except, of course, for measuring disease when we know that you are going to die soon.

What we would like to find are measures of different biological systems whereby measuring these changes and the rates of change in these systems with age, we can get an idea of the underlying processes and mechanisms that are aging, and get an idea to simplify things—whether your aging clock is running faster or slower. We could use that to not so much predict life span, although that would be the first step, but to actually evaluate treatments that are supposed to do something about aging—either retard the rate of aging, or possibly even reverse it for some types of systems.

We haven't been successful in that, but it's a big order and it is a serious question as to whether we will be successful in it.

In order to make the measurements, initially one has to make measurements that don't kill the animals, because in order to study the types of biomarkers that have a possibility of predicting life expectancy, in a mouse, say, we obviously have to keep the mouse alive. In fact, we can't even hurt it. We can't even stress it. We can't even annoy it. We want our mice to be happy and healthy and enjoy having the measurements made.

That puts some limitations on the kinds of measurements that we can make. By making a wide range of measurements we can get a pretty good idea of how systems, like the immune system, extra-cellular macromolecules, wound healing, kidney function, liver function, and to some extent the brain, function, and we can measure a lot of important things.

Brain function, in terms of intellectual function, is hard to measure in mice because their level of intellectual function is so different from ours. But we can measure a lot of important things, and we can follow whether or not specific treatments have beneficial effects on those different biological systems.

That has, I think, a reasonable importance all by itself, even if we don't get to the basic mechanisms of aging. I suspect, Rod or Mort, I don't think there is going to be any disagreement in that we should measure things, and we should try to find treatments that retard or reverse deleterious changes with age.

In some ways, that's what I see as the fundamental use of biomarkers. The more fundamental question is, can we use biomarkers? Can we use total maximum life spans? Can we use species comparisons in order to get at basic clocks underlying all the different changes in the physiological systems and in the pathology resulting, so that we can actually retard, or even reverse, some aspects of aging and increase the healthy period of life span for people?

But first of all, it will have to be for mice.

KONDRACKE: So far, in mouse research, do we have some predictive markers or semi-predictive markers, or what?

HARRISON: We have markers that predict what's going to happen for a population. We don't have markers that predict what's going to happen for an individual.

So, for example, food restriction starting very early in life, and very, very severe food restriction, retards a wide range of aging markers and increases maximum life span.

Dwarf mutations affect the function of the anterior pituitary, so about half the things the pituitary does don't get done and there are little tiny creatures. Those animals have about the same beneficial effects, in terms of life span, as food restriction. Other less severe conditions also have some beneficial effects on maximum life span.

In every one of those cases we have a whole bunch of measures that actually will show that food-restricted animals or dwarf animals or animals with little mutation are aging more slowly. Those are interesting. It's also interesting to ask which ones are consistent with the predictions of the increased maximum life spans and which ones are not.

But within those populations, to have a real predictor—it's not enough to go between these groups because there's a lot of things that happen when you have one of these important mutations, when you have food restriction. Those animals are different in a lot of ways.

So we don't know which of the many ways that we are measuring are actually the ones that are important in causing them to be healthier as they get older.

KONDRACKE: Before we began this, I said if you have to be calorie restricted and you have a mutant gene, and which makes you sterile, the question is whether longevity is worth having!

(Laughter)

—at least in mice. OK, Rod Bronson.

BRONSON: OK. I think that people expect about six things of biomarkers. First of all, it is assumed that whatever change there is, it is going to be linear. So if something changes from the age of twenty to thirty, the same order of change will happen from thirty to forty, and from forty to fifty.

So you can measure any time you want. You can do a five-year study from twenty to twenty-five, or from fifty-five to sixty, and you will be measuring the same thing because the change is linear.

Well, I think that isn't right, because I don't think that aging is linear. I would argue that not much important changes up until about the age of forty in people, and nothing much important changes in mice up until they are about fifteen months.

After the age of forty, that's when everything starts to go to hell, right? You go gray. You become deaf for high tones. Your pulmonary function decreases. Women go through menopause. That's when cancer starts up. Don't worry about cancer before age forty. It's very rare. After forty, the risk goes up exponentially.

So aging is not linear. Therefore, biomarkers are not going to be linear.

Secondly, following that, biomarkers are supposed to sensitive for the short term, as I said. So you are supposed to be able to do a study for five years. You take one group, put them on the fountain of youth extract, and you take the other group and put them on something else, and you run the study for five years, and you get your results. No. In people, aging doesn't happen over a five-year period. It probably happens over ten-year periods, maybe twenty-year periods. So if you are going to create some sort of aging intervention drug or something, you are going to have to run the study for probably ten years, not three or four years. So they are not sensitive in the short term.

Now, it's always assumed that biomarkers are going to be somehow universal, so if you find a biomarker in mice, it will be the same as in rats, same as in cats, dogs, humans, and elephants. I don't think that's true at all. I think that rats and mice age in very different ways.

If you look at their diseases, they are entirely different. For that matter, if you look at the diseases, even within inbred strains of mice, they are quite different. Then you go from mice to cats and there are big changes. If you go from dogs to cats, there are also big differences. If you go from any of those species to people—big differences. So the assumption that you can extrapolate from mice to rats probably isn't true. The idea that you can extrapolate from rats or mice to people may not be true either.

So we may learn nothing about biomarkers. Even if we did find a wonderful biomarker or a series of biomarkers in mice, you still wouldn't know for sure whether they are going to work in people. You still have to do a ten-year study in people looking at the same biomarkers. You are still going to be stuck with a very long study.

Now, there's the assumption that because aging runs by some kind of a clock, or maybe a series of clocks, that all biomarkers are going to go in the same direction and at the same rate. Well, I don't think that's true. If you count gray hairs, the increase in gray hairs such that fifty percent of the people have fifty percent gray hairs by fifty years of age. OK. But the increase in grayness is not necessarily going to be linked at all to bunions. Bunions are also an age-related trait, you know. Young people don't have bunions. You get bunions when you are old. And so on and so forth. So I don't think the biomarkers are going to be linked.

Cancer isn't linked to osteoporosis. Asthma, age-onset asthma, which I happen to have, is not linked to prostate cancer. They are not linked. So therefore, biomarkers shouldn't necessarily be considered to be linked, either. I don't believe they are linked.

As for the non-invasive thing, if you can't be invasive, you are really limited at what you can look at. You can look at body fluids, but not spinal fluid, because no one is going to sign up to get a spinal tap. So it has to be blood. It has to be a limited number of cells. It is unlikely that people will sign up to get a skin biopsy. So if you are going to be noninvasive, you are going to be in big trouble.

Now, in the biomarker study that I was involved in where we studied rats and mice, most everything anybody did was invasive, because you have to be invasive to find inborn changes in cells and so forth. So I don't think we are going to find much unless we can accept a lot of invasiveness.

Then there's the predictive thing. It cannot be true. The ability of the kidney to concentrate urine is well known. Say, for example, you put the person or the mouse on a twenty-four-hour period where they don't drink anything, or a twelve-hour period, and then you collect urine and you look at the specific gravity. A young person will concentrate urine and will be producing very little urine. As you get older, you can't help yourself by dumping urine, because you can't concentrate urine.

Urine concentrating ability can have nothing to do with pulmonary capacity. As you get older, your pulmonary capacity goes down. Your urine concentrating ability can have nothing to do with pulmonary capacity. Your ability to remember ten words, ten minutes later, goes from very good to really bad when you get old. That ability can have nothing to do with pulmonary capacity.

The development of osteoarthritis, which we all get—here it is—there's a little bit of osteoarthritis right there! Here's another little bit right there, right? Everybody gets it. That can have nothing to do with prostate cancer. Nothing. Zero.

So predictiveness is, I think, impossible, unless you are talking about a biomarker of disease. So, sure, cholesterol levels have pretty good, but not too good, predictive value to determine whether you are going to die of heart attack or stroke or something. Even with that example though, there are a lot of people running around with high cholesterol and they never get any disease in their blood vessels. There are other people with low cholesterol who get heart attacks. So even there the predictive value isn't too great. But I am not going to deny that there are disease markers that are predictive for that particular disease, but a biomarker for pulmonary function can have nothing to do with what's going to happen in terms of your brain function or in terms of whether you will get Alzheimer's disease or cancer.

So I don't think that there is any hope of ever finding any useful biomarker.

KONDRACKE: Well, you said that—

BRONSON: Except on a population basis.

KONDRACKE: But gray hairs can have nothing to do with bunions; and urine concentration can have nothing to do with pulmonary function. But are you saying that it is utterly impossible that there could be some sort of common process that goes on where all of these things have various consequences?

In other words, it is some sort of cell breakdown—telomere length or some process that is inherent in genes—that makes you susceptible to disease, slows you down, grays your hair, all of these kinds of things? Are you saying that we will never find one thing or ten things maybe, but there will be an infinite number of things?

BRONSON: Well, that's the basis of the debate here. It actually isn't just about biomarkers. It's about one's view of aging or the theory of aging, which we can come to.

KONDRACKE: No, let's go right there right now.

BRONSON: Hmm?

KONDRACKE: Go ahead.

BRONSON: OK. So there's no question that there are biomarkers of aging. I don't deny it. The test is—look around the room. I can tell that in that corner over there, all those people are young. I never asked them their age.

Now, I don't want to point to anybody else and say that they are, should we say, not young. But it's a piece of cake. Obviously, we all age. For most people, just showing them a picture or having them look around the room, they could peg anybody to within plus or minus ten years. So sure, there are biomarkers.

But are they predictive of anything? Where do they come from? Where does aging come from? That's the real question. Why is it that mice get cancer beginning at about fifteen to eighteen months? People begin to get cancer somewhere between forty and fifty years. What's the difference? How come people live so much longer and how come their biomarkers are delayed compared to mice?

By the way, mice get lots of kinds of cancer. They also get osteoarthritis, and they get all kinds of nasty things, and they always die. They die somewhere between two and two-and-a-half years, depending on the genotype. Same thing with rats. Why is that? Well, you have to go back to this wonderful statement, which nobody understands, but it's the key to the whole thing.

KONDRACKE: It comes from where?

BRONSON: I believe Williams was the one who first said it, and he said that the force of natural selection declines with increasing age. Now, what does that mean? It says that here you are—you're a mouse. You're a studly young guy. You've been producing a bunch of babies and you've been passing on your genes. And you're only, let's say, three months old. So you go out tonight to forage and an owl gets you.

Now, let's say that you are another mouse, and you are nine months old. You go out. You have the same chance as the young mouse of being killed by that hawk or whatever.

Every species is susceptible to accidental death. That is, non-natural causes. Things like drought, fire, flood, murder, predation, and accidents are very important, especially for small animals.

If a male horse decides to mount a female horse and she is not ready for it, she kicks him in the jaw, his jaw cracks, and no matter how studly he was, he is now dead. If you have a broken jaw and you're a horse, you're dead.

So something's going to kill you. If you are a sheep and you are grazing on sandy soil, your teeth will be worn down by the time you are five or six years old, and you will die.

Every species has an upward limit beyond which nobody ever lived, right? For mice it was probably about fifteen to eighteen months. For people, it was probably forty to fifty years. Even now, in the developing countries, very few people live beyond forty or fifty years. Even in Russia, since so much nasty stuff is going on over there, the life span is now about fifty-five—going right back to where it was.

So the reason why the force of natural selection decreases with age is that in any age population there are a few older than younger, because the longer you live, the more chance you have of dying from one of these unnatural causes. In any population there are a lot of studly young guys, but each time they go out and do anything, especially chasing mates while they are not paying a lot of attention to those hawks up there—boom! They're gone!

So as things go on, they disappear. In other words, the young are always going to contribute more genes to the population because they have a better chance of being alive.

KONDRACKE: So this definition doesn't work, obviously. This definition of natural selection decreases with time, basically.

BRONSON: Yes, but the point is that if you get bigger and smarter and wiser, the way people have, you can forage better. You can prevent yourself from being eaten by cave bears. You can maybe get a more orderly life where you are not kill-ing everybody. And now, older people will be living longer and reproducing longer.

So now there is a force of selection of genes to be better buffed up to get you to age forty. As long as you can stave off these unnatural causes of death, there will be strong selective pressure on all your genes to live longer. It's a subtle argument, but it is the key to the whole thing.

KONDRACKE: David? I want to let David respond.

HARRISON: Well, the basic idea that Rod's talking about is the explanation for a very fundamental question that is, why do you age at all? In the simplistic evolutionary theory there's some possibility of living to very, very old age. If you could reproduce at those very old ages, that should give some slight selective advantage. So there should always be an increasing life span, increasing healthy life span, and of course, that doesn't happen. Since we know that's not true, we have to find some way of rationalizing that. When Medawar first thought of the idea—that the ability of natural selection to remove deleterious genes gets lower and lower as the chances of being removed from the population increase, as Rod said.

Williams went one step further and pointed out that not only is it hard to remove the deleterious gene the older you are and the smaller the chance that you have survived, but, suppose that deleterious gene does something good for you early in life—that's the pleiotropic effect. Suppose it actually helps you reproduce when you are very young, or helps you reach the age of reproduction faster and more efficiently, then even though it may have deleterious hard consequences late in life when you are most likely going to be dead anyway, it is not going to be removed. It's going to be selected for.

So those pictures are very, very influential now in explaining why aging occurs. And I think Rod explained very dramatically why there are differences between species.

KONDRACKE: OK.

HARRISON: But there is one other thing that happens that is important to remember, and that is, not only do we all age at rates that are roughly proportional to life spans and life expectancies of our ancestors in the wild, but aging can be changed by natural selection fairly quickly.

There are some studies with fish which suggest that in just a few generations, life spans and natural history rates of development, rates of reproduction, and natural history can change over twenty or thirty-year periods.

KONDRACKE: How does it happen?

HARRISON: Well that's the point. Those creatures must be carrying genes for slower aging.

Now, if aging is caused by an accumulation of thousands and thousands of late-acting deleterious mutations, which is what is predicted by the Williams and Medawar hypotheses, in order to slow down aging perhaps the best experiments were done by Steve Austad with opossums.

KONDRACKE: Tell us about how this illustrated the theory.

HARRISON: Right.

KONDRACE: Because the theory is very abstruse.

HARRISON: Well, the opossums lived on what are now the Georgia Sea Islands. Ten thousand years ago they were part of the mainland. Then when the glacier melted—and we have a pretty good idea of when this happened—the islands were cut off from the mainland.

By chance, the predators disappeared from these islands. So the opossums had much lower predation rates on the Sea Islands than they had on the mainland. That happened maybe five or six thousand years ago. Austad predicted that creatures who are living in a situation with very low predation are going to develop slower, reproduce slower and live longer, and indeed that does seem to be true of the Georgia Sea Island opossums compared to those on the mainland.

That suggests that there are ways of adapting really quickly to conditions that favor increased life span—this is within a species, now.

KONDRACKE: Could be stress!

(Laughter)

Less stress.

HARRISON: Fewer hawks chasing you?

KONDRACKE: Exactly.

HARRISON: Well, probably, if there are fewer hawks chasing you, you wouldn't reproduce more slowly and you wouldn't develop more slowly. You probably would develop, if anything, better with lower levels of stress. So it's probably actually not stress, although that is an innovative suggestion.

It is probably some kind of underlying timing mechanism or mechanisms. Now, why couldn't it simply be reverse mutations of all those hundreds or thousands of deleterious genes? Well, it could. But if you think about what it takes to increase the life span, that requires not just one, but hundreds of things that could have killed you have to not kill you. Right? Because you die of the first thing that kills you. That means that all the things that could have killed you have to not kill you, if you are going to increase the maximum life spans as happened with those opossums.

KONDRACKE: OK. I think we should get off of this philosophical theory. I want to get a little bit more practical.

The National Institute on Aging sponsored an attempt to find biomarkers. Is it now over? How much money was spent on it? Is more investment meritorious or not?

BRONSON: It's supposed to be $20 million that we spent on it. I help spend some of that money.

KONDRACKE: Over how long?

BRONSON: A ten-year period.

KONDRACKE: Ten-year period.

BRONSON: There were anywhere from fifteen to twenty groups over that period of time that looked at all kinds of different parameters—behavior, cellocites, all kinds of things.

We found plenty of biomarkers, as defined by things that change with age, but with a lot of animal-to-animal variability. We certainly found that those things that would change with age changed more slowly in caloric-restricted animals, as we looked at caloric-restricted animals along with full-fat animals.

So in that sense we found plenty of biomarkers. It's just that they are not going to do what people want them to do. That is, they are not going to predict in the single individual. They are only going to work on a population basis.

So I think we were successful. Now, as I said before, they were always invasive and that's not going to work for people. But sure, sure biomarkers exist. It's just that all the things that people want of them are just not going to happen.

KONDRACKE: So, should we keep looking?

BRONSON:—particularly the predictive idea.

KONDRACKE: We should keep looking and spending the money on it?

HARRISON: If you think of one of the critical biomarkers as the age when the female—at least in a mammal—loses the ability to reproduce, that happens relatively early in terms of the total life span in both mice and people. It's roughly half way through the life span that you lose the ability to reproduce.

If you find things that will extend female reproductive life span, for example, what's the chance that that's going to extend total life span? Well, actually, we don't know. But it's probably pretty good.

KONDRACKE: That's the whole hormone replacement therapy question.

HARRISON: Well…

KONDRACKE: No?

HARRISON: Well, hormone replacement therapy isn't extending female reproductive life span. Hormone replacement therapy is trying to prevent osteoporosis after the menopause is over. I'm talking about aging more slowly as far as the female reproductive life span goes. Now, at least in mice, there are genes and there are treatments. One of them is food restriction, which postpones reproductive senescence in females. It also increases maximum life span and seems to slow down a whole bunch of markers of aging.

KONDRACKE: This has come up again and again and again—the idea of food restriction as a way of perpetuating life. Now, what happens when calories are restricted? What does that do to the system, or to the organism as a whole that allows that organism to stay alive longer?

BRONSON: Nobody knows.

KONDRACKE: I see.

HARRISON: Well, it does so many things. Basically, when you are comparing the severely food-restricted animal at, say, eighteen months, with a normal control, it's like comparing a twelve-month-old normal control with a nineteen-month-old. There are so many differences. Probably ninety percent of the things that change with age slow down with age in the food-restricted animal. So it really doesn't help much to answer the fundamental question, "what is the mechanism that does it?" The value of biomarkers, I think we'd all agree, is in trying to understand mechanisms of aging.

KONDRACKE: OK. You know that if a female goes into menopause early that her life span is shorter. But you don't know how much shorter, and it's not a measurable thing?

HARRISON: I'm not sure…

BRONSON: Within a species.

HARRISON: Within a species.

BRONSON: When you go through menopause, that's not predictive.

KONDRACKE: Oh. OK.

HARRISON: I think that that's—

BRONSON: Now, if you breed for increased reproductive life, that might theoretically have an effect on increasing the total length of life. That's what David was saying. But the idea that a woman who goes through menopause at forty is going to die at sixty is just complete baloney. I mean, one hundred percent baloney.

HARRISON: That actually illustrates one of the things I think we all agree on about biomarkers. They are useful in a population basis—we know the food restriction greatly retards female reproductive senescence, or the little mutation which cuts out growth hormone, these things retard female reproductive senescence. But that's on a population basis.

In terms of individuals, there are so many different things involved, especially in a population like humans where you have genetically diverse individuals living in different environments; you really couldn't use a single biomarker, even as evolutionarily powerful a marker as female reproductive senescence, and expect that to make predictions for individuals. That's probably unrealistic. It may well be, however, that if you can make changes, and if you can develop treatment that retard senescence in a variety of biological systems, they will cause the populations in which it's retarded to be healthier later in life.

KONDRACKE: Go ahead.

BRONSON: You could design a very simple experiment. You could take your fountain-of-youth drug, and I would say you should start giving it during that time of life when things really begin to go bad, around age forty. Now, you just take a picture of the person at forty. Then you take a picture of the person at fifty. You do this for a hundred people in the control group and a hundred people in the fountain-of-youth group, and then you take these pictures and you shuffle them up and you give them to a ten-year-old and you say, "Put them in a pile, from older to younger." And they put them in a pile; and we'll find out.

Do you do well during that ten-year period or not? It would work, because just the facial features between forty and fifty are enough of a biomarker. You would know at the end of that period.

Well, let's look at another biomarker—the number of times you go to the doctor for anything other than the normal check-up. If the fountain-of-youth drug is working, you could prove that it is working with that simple experiment. No fancy invasiveness, just take a picture and give them to the ten-year-old and say, "Put them in a pile."

For some people on the fountain-of-youth drug, the forty-year picture and fifty-year picture will be identical. For the people in the control group, the fifty-year-old picture will look older than the forty-year picture. It would work on a population basis. Now, it is not going to say anything about predictors or subsequent longevity.

Maybe the people on the fountain-of-youth drug will all drop dead at sixty. But at least you know that you are delaying aging. You could do it. You're never going to get FDA approval to give anybody a drug for ten years, but you could do it and it would work.

KONDRACKE: Well, there are some products that we had a debate about before that the FDA has no ability to touch that are being sold by the gazillions of pills.

BRONSON: Well, I know. I know, but…

KONDRACKE: But they're utterly untested by the FDA or anybody else.

BRONSON: Yes, exactly. And suckers are buying these ridiculous things all day long. The most important thing that's happened in all of medicine, in all of aging research in the last year or two years, is that female replacement hormones turned out to be bogus. Now, that was approved stuff. This is what all doctors believed was good. Now, there is not a single advantage, except maybe in the short term, if you don't want to have night sweats and all that kind of stuff. But if you take hormone replacement, you are going to die earlier, you are going to get everything earlier. So even that didn't work. It was bogus. So don't take anything.

By the way, there is another paper showing that vitamin E and vitamin C or supplementation also is not good for you. Don't take anything. Take nothing. Buckle up. Quit smoking.

KONDRACKE: Linus Pauling is rolling in his grave. OK.

BRONSON: Oh, Linus.

HARRISON: I think, if we are going to be giving advice to human beings, you have to remember one thing, and that is that the placebo effect is really strong in us. So if you are taking vitamin C and vitamin E, or probably even something as strong as hormone replacement therapy, although that is a little disturbing, but if you truly believe that it's good for you, if you truly believe it, it will be good for you. It will have some beneficial effects.

BRONSON: In the short term.

HARRISON: It may do nothing physiologically, but the fact is the placebo effect is really strong, and that is of course why there is so much junk out there. Because if you believe, if your snake oil salesman is really effective, and you believe in that snake oil, it actually will do some good—not because the snake oil is doing any good but because your belief is doing good, as a placebo effect.

KONDRACKE: OK.

HARRISON: That's one of the things you do have to watch out for, and that's why in aging studies you have to do them double-blind with human beings.

KONDRACKE: OK. Now, what about telomere length? That has been written about as a predictor that as cells reproduce the telomeres—which is the end of a chromosome, if I've got my biology right, shortens and it keeps shortening and shortening and shortening and when it gets to a certain length, the cells die.

BRONSON: There are two things wrong with it: (A) mice have telomerase and they can—

KONDRACKE: Which is—

BRONSON: They don't lose telomeres.

KONDRACKE: Telomerase is what?

BRONSON: It is an enzyme which will stick telomeres back on. People don't have that. So every time a cell divides you lose a telomere.

KONDRACKE: Aging has nothing to do with telomeres?

BRONSON: The second objection, which is a really important one, is that nobody thinks that human beings run out of cell cycles. Nobody believes that. In other words, you have gut cells which are always turning over. Nobody thinks that when you reach seventy suddenly you have no more intestinal cells because they have all lost their telomeres. You don't reach that telomere-absent state where you've lost all your telomeres until there have been more cell cycles than you could ever possibly want.

So I don't know how we got off on the telomere thing. It is just complete foolishness. Don't say anything about telomeres ever again!

(Laughter)

Now, it is true if you knock out telomerase in mice, they do get cancer earlier. It is said that they go gray. Therefore, they are aging like people, because mice don't usually go gray. I mean, who knows? Those studies have to be reproduced in some manner. It may be that there is something there, but boy, oh, boy, their whole company is at the base of the telomere theory. And it's just foolishness. It really is just foolishness.

HARRISON: Well, let's give the audience a bit of an introduction to it. One of the models for aging human beings is fiberglass, OK? Skin fiberglass. Skin fiberglass proliferates in tissue culture and then eventually stops. This was a fundamental discovery made by many people, but actually Len Hayflick was the first one who had the guts to go against the scientific establishment and say that this is what really happens to non-transformed cells.

For a long time people didn't understand why human fiberglass, and sometimes epithelial cells in tissue culture, had limited proliferative capacities. This was used as a model for aging, to try to understand why this happened. And it turns out, although everybody in this field is not completely convinced that it's telomeres, it does seem to be that telomere length best predicts the subsequent proliferative life span of the fiberglass and epithelial cells. That's one piece of evidence.

Another very strong piece of evidence is, if one puts into the cells the essential element for the enzyme to make bigger telomeres—or telomerase—this is, interestingly enough, a reverse transcriptase which came from the AIDS virus. That is how it was recognized.

If one puts that reverse transcriptase, the limiting feature of the telomerase enzyme, into these cells they seem to go on for a very long time, although whether they stay normal for that very long number of proliferations is still argued. Now, the telomerase enzyme uses RNA as a template, so in order to make more telomeres, it matches up the RNA, which is why it needs a reverse transcriptase.

For the mice in which the knock-out was done, they simply removed the DNA, which made the RNA for the template for the telomerase. So the telomeres could not be made any longer. And they bred the mice. The first generation of offspring were fine. I'm talking now just about literature, because I have not been able to get hold of those mice. I tried to get hold of those mice for years and years after those papers were published in top-notch journals, and they never return my calls.

In any case, the literature says, and I believe it, that the mice were fine. They were fine the first three generations. They were fine right up to the fifth generation. Then at that fifth generation the telomeres, incidentally, were getting shorter and shorter. Mice start with very big telomeres. Mice have way bigger telomeres than we have, which is sort of counter-intuitive if you think that telomeres are causing aging, but let's put that aside!

So the mice are actually healthy and, I believe, had pretty normal life spans—at least through three generations—without any telomeres whatsoever. But at that fifth generation both males and females became sterile. That's a little funny, because the number of generations it takes on the sperm line is much more than the number of generations on the egg line. One would have expected the males to have become sterile two or three generations before the females. But, in any case, then all the bad things that Rod was talking about happened.

So that's the history of the telomerase/telomeres. I'd be more hesitant about say-ing that it's absolutely without merit. It's true that few, if any, people die because they run out of proliferative capacity of epithelial cells and fiberglass. But on the other hand, a lot of the aging phenotypes, you know, with skin being wrinkled and so forth—one could argue that the fiberglass is not turning over the collagen properly, and of course, there are other types—

KONDRACKE: But the telomere theory was the holy grail, or the gold stan-dard?

HARRISON: Yes.

KONDRACKE: And it applied to cells all over the body.

HARRISON: Right.

KONDRACKE: And if you could keep telomeres from shortening, you could prevent cancer and heart disease and so on.

HARRISON: Well, actually, by adding the time ratio, you increase the risk of cancer.

KONDRACKE: Ah.

HARRISON: Judy Campisi wrote a neat review a while ago where she talked about sort of a knife-edged balance between cancer and aging. You want to avoid cancer. You also want to minimize aging, so you need a certain amount of proliferative capacity, but if you have too much proliferative capacity it's too easy to transform or fall off into the tumor class.

Just because something doesn't explain everything doesn't mean that we should confine it to the outer darkness. Also, incidentally, I think when people come up with hypotheses, it probably would be wise to exercise a little bit of humility and admit that it's unlikely that all of aging is going to be explained by a single mechanism. Even an optimist like myself.

BRONSON: Well, how about twelve genes, though, right? Three genes? Four genes?

HARRISON: Well, I have no idea.

BRONSON: How about twelve? How about thirty thousand?

You see, when I talk about how people are living beyond the length of time that their ancestors ever lived, my point of view is that it's all just junk. It's not that you are getting new mutations in the genes. Your genes were buffed up to get you through forty, whereas a mouse's genes are buffed up to get them only to about eighteen months.

KONDRACKE: By buffed up, you mean naturally?

BRONSON: Buffed up so they are functioning better.

KONDRACKE: Yeah.

BRONSON: Now, I want somebody to look at obvious issues like cancer, for example. We know that cancer is due to mutations in single body cells and mutations in tumor suppressor genes. We know that there is a lot of DNA repair going on, so if you get a mutation in the gene there are also many other molecules in the cell which will fix mutations in DNA.

Since mice get cancer beginning at fifteen to eighteen months, and people get cancer beginning at forty years, you must have to assume that somehow the tumor suppressor genes in people work better than in mice. By definition they must.

So how about somebody trying to define how it is that we have better tumor suppressor genes than mice? What does it mean to say? If you could figure that out for cancer, because a lot is known about cancer, then maybe you could address the question of why is it that mice, for example, get osteoarthritis beginning at about a year of age? People get osteoarthritis beginning about forty or fifty years old. What's the difference? Well, we don't know the genes of osteoarthritis. But as we learn more about each of these diseases, and if we could ascertain that they happen later in people than in mice, then go after the genes.

KONDRACKE: For specific things.

BRONSON: Yeah.

KONDRACKE: In other words, your theory overall is that aging consists of thousands and thousands of little things.

BRONSON: Yes.

KONDRACKE: Each of which, as we discover them, and if we could fix them for cancer or for arthritis, would contribute to longevity, although it would take an infinite amount of time—so there is no magic key that we're ever going to find to the aging process, or even the ability to measure the aging process.

BRONSON: Well, I would like to start with the first one and answer the cancer riddle that makes mice different from people. Or, if you don't want to do that, I want to know why hair turns gray. I mean, nobody wants to study hair turning gray. You can make a lot of money out of it. Right?

But just take anything and figure out why it doesn't happen until late.

HARRISON: Mice don't get gray.

BRONSON: Why should hair go gray?

KONDRACKE: Well, some people get gray at different rates.

BRONSON: Sure.

KONDRACKE: What is there that is different about people who don't go gray?

BRONSON: Well, it wouldn't hurt to find a gene and clone the genes for early grayness. It may not be the same genes as are involved in all the rest of us, but you could go after it that way.

But even early onset of Alzheimer's, Parkinson's, any disease, just doesn't happen before forty years of age. So what is it? What's the difference? It's not that you have a million mutations beginning at forty. You have the same genes in the same cells, so something happens.

What happens is the genes were not selected to be collectively good enough to keep them from going to pot beginning at forty. In mice that secret age is fifteen to eighteen months. Why are the ages different? Because mice are always eaten up by fifteen to eighteen months, and in the old days people were all eaten up or killed each other off or bopped each other on the head by the age of forty.

KONDRACKE: Well, if you now take the United States of America, or any Western country where the chances of bopping are significantly reduced, the age of the life span is being extended. But it is being extended partly because we've got better medicine. It's partly because we have a healthier life style. It's partly because we are not getting bopped and so on. But there is still a finite limit to the life span for the species.

HARRISON: And a fairly precise one.

KONDRACKE: Yes. The question is, can you either extend the total life span of the species, or at least somehow figure out what it takes to get more of the population to live later into the life span in a healthy way?

BRONSON: Well, if we're lucky there will be maybe a few genes that are responsible for keeping all the other genes buffed up to the age of forty. There may be a few genes. David talked about the opossums on those islands. If it is true that natural selection can increase the life span in just a few thousand years, that would mean that there can't be too many genes.

So maybe there are reproductive age-assurance health genes, and they may be dominant master genes. Now, that's entirely different from saying that there are genes that switch on that make them get old. That is certainly not true.

KONDRACKE: You think there are not.

BRONSON: Zero.

KONDRACKE: Zero.

BRONSON: Aging is not programmed. Aging is the lack of a program. It's the failure of genes to maintain normal homeostasis, to maintain the normal function of—

KONDRACKE: Why does it happen?

BRONSON: The reason why it happens is because there was no natural selection past the age of forty.

HARRISON: No, but that doesn't explain the answer.

BRONSON: Nobody survived.

HARRISON: What the people want to know is why does it happen in a physiological and biological way, not what the evolutionary reason is. I think, actually, we are agreeing that it's possible that there are a small number of genes which tell all these late-acting deleterious genes when it is late. And if that's true, by being able to alter the genes that tell them when it is late, or in Rod's terms, by being able to increase the activity or delay the inactivity of genes that maintain reproductive performance and maintain health, it might be possible to have major beneficial effects on the health span.

KONDRACKE: OK. Questions from the audience: "Many people over one hundred cite positive mental attitude as one of the key factors in longevity. Can you possibly measure mental attitude as you develop biomarkers?

HARRISON: Well, probably just being healthy enough to reach one hundred gets you a real boost. But I suspect there is an intimate feedback between a positive attitude and good health. I suspect that as you get sick, it's harder and harder as you get older and as you have more and more chances of something knocking you down physically—and as you get older and older, this increases. It's harder and harder to maintain a good mental attitude, but there is certainly good reason to maintain it.

As far as a biomarker of aging being a healthy mental attitude—

BRONSON: Sure, you can measure it. There are psychological tests.

HARRISON: There are psychological tests.

KONDRACKE: But is there any correlation?

HARRISON: I don't know.

KONDRACKE: OK.

HARRISON: We both work with mice. And the fact is that mice all have healthy mental attitudes!

KONDRACKE: Not if you starve them, they don't.

HARRISON: Oh, they actually run around and get all excited when you bring them food. They are not starving; you just reduce the amount of food a little bit.

KONDRACKE: I see.

HARRISON: They actually don't have the intelligence, I think, to have the existential questions that cause people to have a really severe depression sometimes. I've never seen a mouse that evidenced the slightest sign of depression.

BRONSON: But do you talk to them very much?

HARRISON: Oh. Of course.

KONDRACKE: OK. "What is the greatest obstacle to effectively discovering a clear biomarker to gauge human longevity?"

HARRISON: That there won't be a single one. I think we can all agree on that. But obviously we need to have a bunch of biomarkers at the very least. The obstacle would be finding the right systems to make the measurements.

BRONSON: And the fact that it would have to be non-invasive, which is a big problem.

HARRISON: Yeah. Although there are clever ways—

BRONSON: There are clever ways, like the ten-year-old girl I was talking about. That would work.

HARRISON: Well, it wouldn't work once they've had the makeovers that they've been offering on TV recently.

(Laughter)

They go from forty to about twenty-five in the picture for the ten-year-old girl. But inside, of course, they are just as rotten as ever.

KONDRACKE: Somebody asked, regarding diet restriction, why is it that people who have anorexia die early?

HARRISON: Well, that's because they don't eat at all! Diet restriction, and I should have said this, diet restriction is not something that you would do with your children. It is severe enough to actually stunt the growth, but also must not be malnutrition. If you get malnutrition, of course, mice will be vulnerable to it and die just like anyone else will.

KONDRACKE: So what percent of their normal calorie intake are you talking about?

BRONSON: Up to forty percent.

HARRISON: Yeah.

BRONSON: And the anorexics will weigh like sixty pounds when they die.

HARRISON: Right.

BRONSON: That's not caloric restriction.

HARRISON: That's starvation.

BRONSON: That's starvation.

HARRISON: There really is a difference.

KONDRACKE: OK. "How will researchers be able to overcome genetic variation among individuals?"

HARRISON: Well, that's a very interesting question, and I think that rather than overcome it, we are going to have to use it. We are going to have to take advantage of genetic variation in the human population and use it to give us hints as to who has the good genes and recognize them.

KONDRACKE: But you can't overcome it.

HARRISON: No. It can't be overcome. It's going to have to be used.

KONDDRACKE: I mean if you have bad genes, gene therapy is never going to work.

BRONSON: Because you can't get genes into all the cells that you need to get them into.

KONDRACKE: You mean you can't use a virus?

HARRISON: Yeah, one can use a virus. Already we have viruses that will target specific tissues.

BRONSON: That will target a few of them. But you've got to get enough virus into enough cells to reverse the genetic problem that you have.

KONDRACKE: Of aging across the board, you could—

HARRISON: You could see the time when we could actually cure something of yours that way.

BRONSON: Well, you could try. OK. Fine.

KONDRACKE: But theoretically—

HARRISON: If you are not too—don't get bopped over the head!

KONDRACKE: But theoretically you could do gene therapies on loads of diseases. Forgetting about the aging processes as a whole, there are specific diseases that researchers are very hopeful about gene therapies for.

BRONSON: Yes. I'm not as hopeful, but yes. And it's worth trying.

Maybe if there are two or three master genes that control aging, which I certainly don't believe, but if there were, then you could measure a person's gene, and if it turns out that the person has a weakness for that particular master gene, you give it some virus or something, and it makes all the genes and all the cells better.

KONDRACKE: Well, that's the golden key theory.

BRONSON: Right.

KONDRACKE: The other theory is that there would be fifty or a hundred principal smaller golden keys and that you could cure Parkinson's, cancer, and heart disease possibly by genetic therapies. But you don't think that's even possible?

BRONSON: Well, people are still going to look a hundred years old when they reach a hundred, even if they haven't had heart disease or cancer. So a lot of the change that you see in people is not going to be reversed by stopping all these diseases.

KONDRACKE: But you would lengthen their life span?

BRONSON: Yes.

HARRISON: A healthy life span.

BRONSON: And that might be worthwhile.

HARRISON: It might be that appearance might be changed, too, although that wouldn't be changed by curing the specific diseases. There's a lot that we don't know about how many critical timing mechanisms there are that cause aging to be thirty times faster in mice or seven times faster in dogs than in humans.

But there are methods already developing to begin to look for those genes, and species comparisons, massive species comparisons. There are some neat critters—bats, for example, bull rats, colonial organisms, that live way longer than other creatures of their size and metabolic rate.

KONDRACKE: Why?

HARRISON: It would be interesting to find out why.

BRONSON: Why do zebra fish that are this long live for three years?—a lot longer than mice. There must be a lot of predation for a little guy like that.

KONDRACKE: "Does something need to be discovered before a defined biomarker can be established or, are biomarkers necessary to define what aging is and what researchers should be looking for?"

HARRISON: I'd vote for the latter.

BRONSON: Yeah.

HARRISON: I mean, we have accepted the definition of biomarkers, now we are talking about measures of aging rates in a wide variety of biological systems in individuals. I think we agree that it's necessary to follow those as much as you can when you are studying aging, whether your focus is on carrying specific diseases or whether your focus is on trying to do something more basic—or, what is really logical, to go after both.

BRONSON: I think probably if we found real biomarkers, we would know a lot more about mechanisms of aging.

HARRISON: Because they would be following the specific mechanisms.

BRONSON: Yes.

HARRISON: It would be nice if, as we define specific mechanisms of aging, we could also define their biomarkers. But something like the telomere, that illustrates the problems that can happen. You find one thing that seems really good in one particular model system and you say, "Ah, it's going to cure everything." Then it turns out not to. And then, of course, people say, "Well, it's no good at all."

We are probably going to need a number of different measures. I'm not absolutely certain that telomeres aren't going to prove to be important.

BRONSON: A little bit, sure.

KONDRACKE: "If you were advising Congress of where to put its money, would you advise Congress to spend the money on curing specific diseases or for trying to discover what the aging process is all about?"

BRONSON: Well, they've got to cure the diseases, for sure.

HARRISON: Politically, you couldn't advise the other.

KONDRACKE: Well, forget about the politics.

HARRISON: We can beg them to spend a little bit of money on trying to find basic mechanisms.

BRONSON: That's right.

HARRISON: That's the best we can do.

BRONSON: And it's got to be mechanism, mechanism, mechanism. The problem with the biomarkers thing is we weren't even asked to go after mechanism. I had no idea what the mechanism was. I still don't.

But it must be mechanism, mechanism, mechanism. What is the mechanism of aging? So for sure, the government should not fund pure biomarker research. Forget it. Fund research where you find biomarkers that are reflecting a mechanism that you're studying, that's fine—justified biomarkers that reflect the black box of aging. We shouldn't do that again. We've got to get mechanism.

KONDRACKE: I'll give you the final word. Is that right?

HARRISON: It's necessary to use biomarkers to find mechanism, we agree. It certainly doesn't hurt to have one person studying biomarkers and finding out what the best biomarkers are so that the rest of us can use those to find mechanisms.

KONDRACKE: OK. With that we will thank everybody for being here, and I think we've had a great discussion. Thanks a lot.

End.

How Soon Until We Control Aging?

Aubrey de Grey, Cambridge University
Richard Sprott, Ellison Medical Foundation
Morton Kondracke, Moderator
November 5, 2003

Pictured: Aubrey de Grey, Cambridge University.

For more information on debate participants and SAGE Crossroads go to
www.sagecrossroads.net

KONDRACKE: On my left is Dr. Aubrey de Grey, who has achieved an enormous amount of publicity lately for predictions that I will let him make in detail, but who basically claims that we can live forever. And soon. He is at Cambridge University's Department of Genetics in England. His opponent in this debate is Dr. Richard Sprott, who is the executive director of the Ellison Medical Foundation.

DE GREY: Thank you. Should I go to the podium?

KONDRACKE: No. You can stay here.

DE GREY: OK. So, I'm going to give a very specific numerical statement about my feelings about the time scales for serious progress in doing something about human aging. I am going to suggest that by 2030, twenty-seven years from now, we will probably have the technology to take middle-aged people, people aged fifty, shall we say, who are typical fifty-year-olds, in reasonable health, and give them fifty extra years of healthy life expectancy over and above what they have today.

So today, a fifty-year-old might be expecting to live until they are eighty. I'm saying that by 2030 we will have the technology to get them to live to about 130. And those extra years will be healthy years. That's very important not to forget—that this will not be an extension of frail life.

OK. That's the case I'm going to make, and I am going to make it in the most direct way possible. I'm going to tell you how we can do it.

I'm going to set out a panel of interventions, which I believe we can implement in that timeframe, and I am going to set them out in sufficient detail to justify my confidence that we can do it in that time frame.

I want to start by defining aging in a way that helps me explain what we need to do in order to affect aging. There are, of course, many ways to define aging and living long, good years.

Aging is a side effect of being alive in the first place. So being alive—the whole myriad of molecular and cellular processes that keep us more or less the same from one year to the next. It is an incredibly complicated network of interacting processes. I am going to just use the word metabolism to cover that whole thing.

Now, if metabolism were perfect, in other words, if it really kept just completely unchanging from one year to the next, then we wouldn't have aging, because, obviously, the remaining life expectancy of an organism as a result of intrinsic mortality is a consequence of something that's actually different in an older person relative to a younger person.

So something must go wrong. And indeed, what goes wrong is that metabolism has side effects. And those side effects accumulate. At the molecular level and the cellular level, there are microscopic changes that accumulate in our bodies over time.

At first those changes are harmless. They do not have an effect on the function of our tissues and on our susceptibility to age-related diseases and so on. But eventually they reach a level of abundance that is bad for us. It's pathogenic. So it causes aging-related diseases and eventually it kills us.

But the critical thing to point out here is that that only happens once these types of damage have reached a certain level of abundance. So a thirty-year-old, for example, is more or less in as good condition in terms of how they work, as a twenty-year-old. But we know there is something subtly different about the thirty-year-old because the thirty-year-old has a shorter remaining life expectancy.

OK. Now, why do I want to define aging that way? First of all, I should point out that what I've said so far isn't really at all controversial. I very much doubt that Dick will have any difficulty with that definition of aging for the purposes of today.

The reason it is useful is because it tells us what we can do that we might not have thought of. When most biologists are pessimistic about the likelihood of doing anything serious about aging any time soon, the reason they are pessimistic, in my view, is that they overlook the alternatives that are available to us.

Biologists tend to look at evolution as a good example of how to fix aging, because evolution has very successfully taken a short-lived organism and turned it into a long-lived organism on many occasions.

But evolution has very, very different tools than we have. Evolution can only work with the products of spontaneous mutation and such that occur in organisms that can be selected. We have the tools of molecular biology and cell biology in addition.

If we only had the same tools as evolution, and if all that we could do to aging was to follow the same sorts of strategy that evolution has followed, then I would completely agree with all of the pessimists in this building, including Dick, that it would be many, many decades before we could do anything half as much as I am saying we can do by 2030.

The reason we can do better, in my view, is because we have an alternative. Evolution works basically by making metabolism cleaner. So I've defined aging as, essentially, metabolism causes damage and damage eventually causes pathology, which kills us. Evolution slows down the rate at which metabolism causes damage, so it delays the point at which damage gets to a level that caused pathology.

Fine.

We have the alternative of focusing not only on metabolism, but on the damage itself. We can let metabolism do the stuff it does, lay down its damage, at the rate that it naturally lays it down, but we can also go in periodically and repair that damage—actually fix it so that it's never allowed to get to a level of abundance that causes pathology. And by that means we can unlink metabolism, or being alive, from pathology, being dead.

Why is this easier? Well, basically, the reason why it is easier is because of complexity. Metabolism is unbelievably complex. It's a real mess. We have to understand it well in order to mess with it, and that's why it's very hard to mess with it and we won't be able to do so effectively for a long time.

The pathology that we eventually die of is also very complicated and messy, certainly. No question. But the linkage, the damage, the microscopic damage that links the two, and without which there would be no pathology from age-related causes, is simple. There are really only seven major types of damage that actually accumulate during time, and if we could fix them all we simply wouldn't age.

These seven types are as follows. I am going to go through them very quickly, and I am going to suggest to you the essence of how we can fix each of them.

First of all, there is cell loss. Certain tissues of ours lose cells with time, and those cells are not naturally replaced, so certain areas of the brain, of course, and also the heart, are examples. This is what stem cell therapy is for. Stem cell therapy is an area which is proceeding really quite nicely at the moment, despite the regulatory obstacles that it faces in some countries, and it's working pretty well to augment our natural regenerative capacity in those tissues that don't have enough to start with.

So it's not particularly outrageous to suggest that by 2030, twenty-seven years from now, we will have cell loss pretty much licked.

Second is mutations in our chromosomes. Mutations in our chromosomes cause cancer, and there is very good reason to believe that that's all they cause that matters to us in anything like a normal life span. It is so important not to die young of cancer that we have really, really good DNA repair that gives all the other genes that have nothing to do with cancer a sort of free ride.

Therefore, all we have to do in order to eliminate mutations in our chromosomes, as relevant to aging, is to cure cancer. Now, that of course, is easier said than done. We've been trying for a while with only modest success. But stem cell therapy gives us a massive new opportunity that we have never had before.

The gist of this new therapy is to replace stem cells progressively and repeatedly in all our tissues with ones that do not have a gene for a special enzyme called telomerase that maintains the ends of chromosomes and allows cells to divide indefinitely.

So again, the hard part of this is stem cell therapy, which is going pretty well already.

The third thing we have to fix is what are called mitochondrial mutations. Mitochondria are special machines in the cell that have their own DNA. They only encode thirteen proteins of their own. All the other proteins that make up the very complicated mitochondria come from the nucleus. So the way to obviate, rather than eliminate, mitochondrial mutations with respect to aging is to make copies of these genes and put them in the nucleus, suitably modified so that they still work.

That sounds pretty damned ambitious, just said boldly like that. Right up until you hear that it was actually done successfully fifteen years ago—only for one of those thirteen genes, and only in yeast, but it was done and there has been more progress in mammalian cells with two more of those proteins in the past couple of years. So we are moving fast here.

Number four is cells that we have too many of and that don't do any good. We should have gotten rid of them, but they hang out even though we don't want them to. Senescent cells are an example of this. Senescent cells are normally studied in the cell culture dish, but they also occur in the body and we really ought to just get rid of them.

And then the major example is fat cells, especially in the abdomen, which cause insulin resistance and diabetes. These are cells that we need to get rid of, and work is proceeding very successfully in model organisms to make those cells commit suicide or, alternatively, to activate the immune system against them so they are destroyed.

This is work going on in the labs of senior, respected gerontologists, so it's proper work.

Number five is extracellular cross-linking. This is the thing that causes hardening of the arteries. Basically, it's a chemical reaction between sugar in the circulation and long-lived proteins that make up the artery wall and other tissues that are long lived.

We are lucky here because it turns out that the chemical structure of these cross links is different from anything that we make naturally. So it's been possible to design drugs that can break the cross links without serious side effects. One of these drugs is already in clinical trials, which is a measure of how far along we are.

Number six is extra cellular junk, or garbage. This is most important in Alzheimer's disease, where you have big aggregates of a special protein called amyloid beta that accumulates between cells, and again, we have a therapy that is already in clinical trials, not simply to slow down the accumulation of this stuff, but actually to get rid of it after it's been laid down by activating the immune system against it.

That's worked because once this junk gets inside cells, the cells can apply more machinery. They have special machines called lysosomes, which do this.

That brings me to number seven, the last one, which is the breaking down of junk that accumulates inside cells. Junk accumulates because there are some things that even the lysosome can't break down. There the most promising work is actually going on in my department in Cambridge, in England, where we are identifying enzymes and genes for enzymes from the soil from microbes, bacteria and fungi in the soil, which can break down the things that we can't.

These turn out to be easy to find. No surprise, because, of course, when we are buried, these things are degraded even though we couldn't degrade them ourselves when we were alive.

So those are the seven things, and that gives you a feel for how hard and how fast the science to fix these things is already going. That makes me feel that we ought to be able to develop all of these technologies in mice within ten years and, with the impact on expectations for biomedical progress that that will cause, it seems very plausible to me, not remotely outrageous, to suggest that within another seventeen years, by 2030, we'll be done. We'll have transferred those technologies to humans.

Now I want to stress, before I close, that the ideas I just described are not just my ideas. They have been extensively scrutinized by senior experimental scientists and have not been found glaringly wanting. The first publication of all this stuff that I made was about eighteen months ago now in conjunction with a bunch of very famous gerontologists, people like Bruce Ames, Judy Campisi, Roger McCarter. They signed up to this because they felt that this was not sufficiently implausible. And it's been published and it's still there. So it's not science fiction.

So, to conclude, I'll just say that in summary I have given eight separate ways to show that my optimism is misguided. I've give the possibility to identify one of my seven areas of aging that is actually much harder to fix than I say, and therefore we won't get anywhere to speak of with it by 2030. Or, you can identify an additional component of aging, which would still continue, and still kill us, even if we did fix all the seven things I mentioned.

So I'll stop there and I'll let him do his worst. Thank you very much.

KONDRACKE: Well, before we get to Dr. Sprott, and as a kick-off to Dr. Sprott, I would say that you sort of understated your optimism here. I've seen you quoted as saying that people who are alive now will live to be one thousand years old, and you've also been quoted as saying that by the year 2100, people will be living four thousand to five thousand years. Do you stand by those predictions?

DE GREY: Sure. Can I answer that numerically?

KONDRACKE: Yeah. Sure.

DE GREY: OK. So actually your mistake there is to suggest that the predictions you just mentioned are more optimistic than the prediction I've just mentioned. In fact, if my prediction actually comes to pass it's more or less a given that we will have people already alive that will live to a thousand years and people born in 2100 will live even longer, because of the bootstrapping.

If you think about it, what I am talking about here is rejuvenation. Not slowing down aging, but actually fixing people up who have already aged somewhat. So supposing we are at a point where we can take, let's say, a fifty-year-old and make them live fifty years longer than normal.

Now, fifty years is an eternity in science, so that person is going to be around for the next generation of antiaging treatments that are cleverer and can fix up someone who is even older, right?

So really, we will have reached easily escape velocity by that point, you might think of it that way, whereby it gets progressively easier to keep people going, even though they are older, because we are finding out and fixing the things that go wrong with us faster than we are encountering them.

KONDRACKE: Later we will get into the philosophical discussion of why anyone would want to live five thousand years, but let's go to Dr. Sprott.

SPROTT: And I didn't pay Morton to ask that question!

I apparently got myself into this debate as a result of my comments after the Greg Stock-Bill McKibben debate, "Do We Want Science to Reinvent Human Aging?"

Much of that debate was devoted to the promise and premise, both technical and ethical, of germ line engineering, which includes gene therapy and related technologies that could be used as interventions that would retard or even reverse aspects of aging.

My comment at the end of that debate was essentially that I am not convinced that aging is a disease or genetically programmed in ways that would allow us to attack it in the same way that we attack diseases.

Given that conviction, I think that one of the real dangers of the push to control aging by scientists who are interacting with the media and seeking specific kinds of legislation to enable or prevent such developments is that we risk wasting very precious fiscal, intellectual, and political resources on what I think is a highly unlikely goal, at the possible cost of not pursuing research on the diseases of aging, which would in fact improve the lives of nearly everyone on the planet.

The title of this debate, which Aubrey and I agreed on, was "How Soon Will We Be Able to Control Aging?" I think it makes the rather astounding assumption that control of aging is a given, and that the only question is how long it will take us to get there.

As Aubrey's already stated, he and I agreed that for the purposes of this debate we would assume the control of aging means that by the year 2030 we will be able to take people aged fifty, in generally good health and hence, with about a thirty year remaining life expectancy, and extend their healthy life span by fifty years, on average, even without using any therapies not yet invented by 2030—that's bootstrapping.

DE GREY: Yes.

SPROTT: Thank you. And that would produce human beings with a healthy life span of 130 years.

Since the longest lived human so far, Madame Jeanne Calment lived only the age of 122, raising average life expectancy to 130 years would be quite a feat. But it only begins to approach the sort of promises that are being bandied about in the media, and by a few of our scientific colleagues. One hundred and fifty years is a commonly used figure, and it's a lot less than 5,000. And I think it's even more unlikely that we get there.

However, let's stick to the 130 years that Aubrey and I agreed we would stick with for this debate.

Aubrey's argument, as I understand it, both from his comments here this morning and his written exposition of his views in other places, rest on, I think, a few main assumptions.

He enumerated eight assumptions, and I am not going to deal with all eight because I don't think we need to. They rest on a few assumptions, and the logical consequences of those assumptions as he sees them.

Assumption one is that the originating cause of aging is clear. It results from our being alive in the first place. I can't argue with that.

The idea that if we could not age, we would not exist is cute, but I don't think it's terribly useful for this particular discussion.

More to the point, and Aubrey does make this point, being alive requires maintenance of the organism. Damage to the system results from the processes needed to keep the organism alive and functioning, and from the slings and arrows hurled by the environment. Even in the absence of an aging process, we wouldn't last forever because there are those trucks out there aimed at us, or what have you.

In fact, most definitions of aging start with these simple facts and assume that aging is the loss of the ability to repair that damage.

The definitions vary in sophistication and complexity, and I don't think they need to be debated here. But most gerontologists, I think, do subscribe to some version of that definition.

The other possibility is, I think, that aging is programmed in our genes just like early developmental events. At the species level this must be true. It's not an accident. I think that guppies live to be about a year, maybe; dogs, 7 to 15 years; chimps, 30 to 50 years; and humans up to, so far, 122 years.

The longevity program, however, is likely the unintended consequence of selection for early life reproductive factors, not selection for greater age.

The variance that we are really most concerned about here today is not that species variance, but the variance within a species. We want to understand and control the factors that make it possible for some human beings to live somewhere between eighty and a hundred years in relatively good health, and others get only half of that amount.

What Aubrey wants to do is to add another fifty years to that expectation. I think that's a whole different kettle of fish.

His second assumption, as I understand it, is that the best approach—and this is the key assumption of his point of view—is that the best approach to achieving this increased life expectancy is to prevent damage from overwhelming the organism—that's you and me—by repairing the damage periodically, rather than by preventing it in the first place.

I would certainly agree that preventing the damage is not possible, but I'm not even vaguely convinced that we could repair the damage periodically with any real success.

Len Hayflick, who is well known for his views on this topic, uses the auto repair analogy, pointing out that we are not able to keep an organism as simple as an automobile, which is simple compared to us, functioning for a life span of anything like a hundred years without shoving it in a museum, thus removing it from life. And this is with an endless possible number of replacement parts that can be plugged into that organism.

The notion that we know or will know enough to control an organism as complex as a human being well enough to accomplish this seems to me to be arrogant in the extreme. Granting that molecular biologists don't lack for intellectual arrogance does not grant that they know enough to accomplish that task.

Now, before I get shot by all my molecular biologists friends, I have to point out that they are mostly very, very nice people. They are engaged in very worthwhile research that will surely have impact on the human condition.

What many of those investigators are attempting is to alter the human genome in ways that will prevent the damage from occurring, or that will improve the efficiency of innate repair mechanisms.

I simply don't think that's going to happen any time soon, and I don't think Aubrey thinks that's going to happen any time soon, either, since he's chosen another route to go down.

Indeed, I don't really think it's going to happen at all, but that's the subject and part of this debate.

The example of hormone replacement therapy, I think, is very instructive at this point. We've been pursuing this approach to preventing or repairing the damage caused by age-related hormonal decline for nearly fifty years, and we clearly don't have it right yet.

I think this is an eloquent demonstration that the human organism is enormously complex, and we don't know enough to override our genetic heritage. To assume that we do know enough is, at the very least, overreaching our demonstrated ability and, at the worst, possibly quite dangerous.

Aubrey obviously disagrees. He has stated elsewhere, although not this morning, that the endocrine system is a relatively straightforward system to repair if such repair is needed.

The endocrine system, however, I think, is just one example of the systems that decline with advancing age. The immune system, the cardiovascular system, the central nervous system, and so on, all decline, and each would need to be successfully repaired.

This would require repair capacity that boggles my mind, and probably most of yours. Unless you've been extraordinarily lucky, your experience in the auto repair shop for humans—otherwise known as the HMO, or, if I want to be a little more fair, some of the nation's finest hospitals—probably doesn't give you much hope that this is a doable thing.

The other approach would be to find the master gene or genes—what we will call "gerontogenes"—that control all age-related changes that lead to our demise and to re-engineer those genes.

That way, the "gerontogenes" would take care of all of the difficult work, coordinating the very complex tasks that need to be accomplished to produce significant life span extension.

It's not the approach that Aubrey favors, as he stated elsewhere, that the complexities involved are not likely to be overcome any time soon.

What he does advocate was summed up in his January SAGE article; first using gene therapy—and he did it again, by the way, this morning, the same points. First, using gene therapy, we would need to enable cells to break down intracellular aggregates, what he calls junk, by giving those cells extra enzymes that can degrade the junk.

These enzymes are not yet identified, but Aubrey is hopeful that they will be by the time we have the gene therapy techniques in hand that will allow their insertion into every cell in the human body.

Next, again, using not yet developed gene therapy techniques, we would need to make, and I quote, "Fairly obvious changes in DNA sequences that encode important mitochondrial proteins, and then put that DNA into the nucleus of cells." And he told us more about that a few minutes ago.

As Aubrey has pointed out, these proteins are very important. Damage to mitochondrial DNA may well be a key part of aging decline, and Aubrey's approach would be to make mitochondrial DNA superfluous. The fairly obvious changes in DNA sequences would accomplish that task.

Here, too, the changes would need to be engineered in such a way as to affect every important cell in the body. Then, because of the very major role that cancer plays in aging decline, we would need to eliminate it in all of its forms.

This would be accomplished by "total elimination of the genes for maintaining telomere length from all mitotic cells." I think that even if this were technologically feasible it rests on an oversimplified view of the role of telomeres and telomerase in cell function in cancer. It also assumes that telomeres serve no useful purpose in a mature organism, and I think that's an iffy proposition, as well.

In order, then, to produce the life span extension that Aubrey asserts is possible by 2030, we would need to achieve not just one, but every one of these redesigns of the human genome and repairs, and I submit that that's simply not going to happen.

While I recognize that my assertion is every bit as much a statement of belief as Aubrey's, I think that probability is on my side. So, if there is a little time here, what do I think is possible and desirable?

I'm not proposing any great change in the research that we are conducting. The biology of aging research currently funded by the NIA, by the Ellison Medical Foundation, by the American Federation for Aging Research, by the Alliance for Aging Research, can have enormous impact on the health and well-being of humans world wide. It doesn't really matter why an investigator is motivated to seek greater understanding of the basics of biological aging, so long as that research is honest and honestly reported.

What does matter a great deal, however, I think is how that research is presented—read "sold"—to the public and to Congress. In both instances, hype about producing significant increases in human life span produces unreasonable expectations that are bound to be disappointed.

We have all seen the growing disenchantment of the public, the Congress, the media, ourselves, with the continual release of recommendations and warnings about health effects, good and bad, of substances from common foods like coffee, to treatments like hormone replacement therapy.

A great deal of public good will and energy is wasted on false hopes created by well-meaning scientists. The fact that the quacks and the hucksters trade on those hopes and the ammunition that we hand them just makes the situation, I think, all the more difficult.

What we can do is take the high road and refuse to hype great increases in life span for fun and profit. I'm not really naïve enough to believe that those biogerontologists with fiscal interests in the private sector, that are based on the promise of increasing life span, will suddenly quit saying this stuff in public. But I think the rest of us can provide a tempering voice whenever we are given an opportunity to do so.

In the long run, I think the dividend will be greater support from the public and from Congress and from our fellow scientists. I really do believe, as well, that most people over the age of sixty-five or so aren't really all that interested in adding another fifty years of decline to their lives.

What they are interested in is making the last third of their lives healthy and productive. Producing that outcome would be a fine achievement, and I think a fitting payoff from the new biology that was paid for by that same public.

KONDRACKE: Thank you, Dr. Sprott.

Clearly there have been advances in longevity and life span over the last decades. What do you think is possible in terms of lengthening average life span over the period from now until 2030? That's only twenty-seven years.

SPROTT: OK. And there are two parts to what you just asked, and that's average life span versus maximum life span. Madame Jeanne Calment is an illustration of the known maximum human life span. Average life span is considerably less than that, and what we've seen throughout most of the last century is increases in that life span that resulted from eliminating infectious diseases, better hygiene and better maintenance of the organism.

One of the interesting questions is whether the kinds of changes that we see in mice and rats with caloric restriction, which by the way, took place also in the context of very, very different husbandry, would be repeated in the human situation where we have pretty much already maximized the environment.

So those are interesting kinds of questions.

What do I really think is possible? I think we might see average human life span approach the mid-nineties, another decade beyond where we are now.

KONDRACKE: And that's fundamentally by curing the diseases of aging?

SPROTT: Yes. I think everybody here is probably seeing the basic thing: if we eliminate cancer, heart disease, diabetes, obesity and most of the things that kill people now, we would add about seventeen years to human life expectancy, which would get us a little beyond what I just said.

KONDRACKE: OK. Now, do you see in any of the seven areas that Dr. de Grey mentioned sufficient progress that major advances could be made in any of them in the timeframe that he is talking about?

SPROTT: I think it's possible that we might see some genetic reengineering that will have a major impact on our susceptibility to some diseases. It could have an impact on our ability to repair some kinds of damage. Where I disagree is I think you have to do that whole list that Aubrey presented or the vast majority of it to get there, and that's what I don't think is terribly likely to happen in the near future.

KONDRACKE: Your chance to rebut.

DE GREY: OK. Well, the first thing I want to mention that Dick said, which I would like to dispute, is his assertion that his view is a statement of belief almost as much as mine is. I would say that it's a statement of belief much more than mine is, because what Dick hasn't done is shown me the miracle.

Those of you who know what I am talking about will remember a very famous cartoon that was produced where two professors are discussing some particular problem by a blackboard, and one of them has claimed to have solved some important problem.

And he's written a large amount of algebra on the left and a large amount of algebra on the right, with arrows between them. And in the middle it says, "Then a miracle occurs." And the other professor says, "I think you need more detail here in step two."

Dick hasn't taken on any of my eight challenges to him. He has not identified a specific molecular or cellular mechanism that might contribute to aging, let alone does contribute to aging, that is independent of the ones I mentioned. He also has gone only a very small way to identifying any difficulties that exceed the ones I suggested in implementing my proposed solutions to each of those seven strands.

One thing he said about cars was interesting. It's actually perfectly possible these days to keep a car on the road that's a hundred years old. And what's rather interesting to observe in vintage car races and so on is that it's not actually all that much more difficult to keep a car on the road for a hundred years than for fifty years. You are more or less fixing everything by the time you get a car out to fifty years. It's hard work to do it. I'm not saying it isn't hard work, but it's not beyond us.

The point that Dick made about hormone replacement therapy is an important one with regard to the science of all this.

Hormone replacement therapy indeed is something we are not terribly good at yet, which we ought to be better at. But the reason it is so difficult is precisely the reason I gave.

Hormone replacement therapy is an attempt to mess with metabolism. It's an attempt to mess with short-lived, bioactive molecules that are circulating in our bodies. Short-lived molecules, by definition, cannot be components of what I have called "damage" because they cannot accumulate. They are broken down or excreted or whatever. And there may be a higher abundance of those damaged, short-lived molecules in an older person than in a younger person, but logically that has to be a consequence of the accumulation of damage in a long-lived way, either in long-lived molecules or in long-lived tissues that are not replacing their cells, for example. It cannot be the primary cause of anything.

So what I am saying is that you wouldn't need hormone replacement therapy if we fixed all seven things that I described earlier.

Now, another thing Dick mentioned was the research agenda. I think it is indeed correct that it will be a very long time before it becomes superfluous to carry on doing research to understand aging better. That's obvious. That's not where we differ.

But historically, there is a strong tendency, when a field with technological or biomedical relevance moves forward enough, for those scientists who actually made the critical progress in our understanding that made the solutions possible actually not to realize how close they were to solutions.

And you can think of cases, like, for example, the Wright brothers going ahead and building planes just three weeks after some professor at Case Western published a proof that heavier-than-air flight wasn't possible. And this is just one example of many people who had said flight wasn't possible right up until it was done.

It's important to remember that people have specializations. People have different ways of thinking. The engineering type of creativity is in many ways very different from the basic science type of creativity.

The last thing I want to mention is that the attitude of the public to perhaps less than predicted rate of progress in an important field that costs a lot of money is definitely important to consider, but we have a much better precedent for that than any that have been mentioned so far, which is the war on cancer.

Cancer research has progressed well over the last thirty years, but it has progressed far less well than Nixon thought it would in 1971 when he announced the war on cancer and announced this big hike in the budget of the National Cancer Institute and so on. Has that turned into disillusionment from the public or a diminution of funding? You bet it hasn't. The NCI is still funded in an increasing amount every year. When the public gets its teeth into wanting to fix something, the public will accept that science is not predictable, but that we have to try our best.

KONDRACKE: Now, what about his point that you have got to succeed not just in one of these areas but all seven? You've got to run the gauntlet here and do it all.

DE GREY: Sure. I didn't mention that because I absolutely agree. It may be we can get away with five of them. I think mitochondrial mutations and senescent cells are two areas in which the jury is still out as to whether they really matter in anything like a normal life span.

But they might matter, and all the others more or less definitely do matter. So I am basically with Dick on that one.

KONDRACKE: OK. I want to remind the Web audience that they have about twenty-five minutes to send in questions. The instructions are on the screen.

Now, you yourself acknowledge that this is really complex and really messy. And yet you think that all this can be done in seventeen years?

DE GREY: That's not quite what I said. I said that metabolism is really complex and messy and the pathology is really complex and messy, but the damage that links them together is not complicated and messy—it's still fairly complicated, but the degree of complexity is massively, massively less and that's why it's within range.

KONDRACKE: Oh. OK. So now just considering brain disease itself. I mean, here you have this enormously complex structure called the brain. And I happen to know something about—

[BREAK IN TAPE]

—of any of them because we are going to cure it in fifteen years. That was fifteen years ago and we are still working on endless numbers of therapies to try to do that. Stem cells offer a hope of possibility of that. But it's not at all clear that that's going to happen.

So why are you so confident that in just twenty-seven years we will be able to replace neurons of any kind with stem cells?

DE GREY: OK. First of all, I'm not saying we're definitely going to be able to do this. I am only saying we are probably going to be able to do this, OK?

So clearly these things might take a hundred years. We just don't know until we've done it. But I said the probability is better than fifty percent.

The second thing is, I mean, stem cell therapy is moving really fast. It's something that wasn't moving so fast fifteen years ago. Maybe that fifteen-year prediction was over-optimistic. Maybe my prediction is a bit overoptimistic. Maybe it's a lot overoptimistic. We are definitely not at a position where we can carry on not even trying.

KONDRACKE: Well, there's no question that we should be trying. But what is your forecast on what stem cells can do to solve this part of the problem?

SPROTT: I think what stem cell research can offer us is the opportunity to repair and replace cells in a very different way than we currently do. The possibility that we could, in fact, provide replacement body parts through stem cell replacement—I think that's a reasonable possibility. Whether I'd say we could do it in twenty-five years or fifty years, I don't really know. I don't think we are going to do it next year.

DE GREY: One down, six to go.

SPROTT: If the UN has its way, we are not going to do it ever, but—

KONDRACKE: What's happening at the United Nations is that the Bush administration is trying to get the United Nations to discourage cloning of embryos for stem cell research. But even if it succeeded, the UN can't stop Israel or England or anywhere else from doing that, just to be clear on that.

OK. Go ahead. Continue.

SPROTT: Well, I had mostly gotten through that.

KONDRACKE: Now, as to replacing cells that produce telomeres, your second area of success. Various aging experts that have been here say that what we want to do is produce telomeres in order to prevent the shortening of telomeres in chromosomes. If I understand this correctly, it sounds like you are suggesting the exact opposite as the answer to the aging problem. Would you explain that please?

DE GREY: Sure, and this is actually not terribly controversial within gerontology. It's actually extremely rare to find anyone who knows really anything about this biology who thinks that giving yourself telomeres will be good for us.

Telomerase is an enzyme that's in all our cells, and in most of our cells it's very, very robustly turned off. The general consensus in the field is that this is precisely because telomerase allows cancers to kill us. OK. So telomere suppression has evolved as an anticancer mechanism that makes us live longer because we don't die of cancer.

The idea of turning off telomeres is not just by turning the gene off, but by removing it so that even the very hyperactive mutability that exists in cancer cells will not be sufficient to restore the ability of the cell to divide indefinitely its telomere shortening.

And, of course, some cells in all of our tissues that are rapidly renewing actually need telomeres in order to divide often enough so they would cease to function as stem cells after a while, but we've good reason to be believe that we would only need to replace them with new cells that again lack telomeres, but had nicely refreshed long telomeres that were refreshed outside the body every ten years or so. Going in for a ten year service is a small price to pay for an indefinite avoidance of cancer.

KONDRACKE:—body wide that you have to eliminate telomeres.

DE GREY: Body wide, but with different types of treatment in the details for different tissues. So, for example, in the blood it's relatively straightforward—a bone marrow transplant is all we are talking about here, and that's a relatively routine therapy already.

In the skin, for example, we are talking about replacing the stem cells that keep the epidermis going, the outer layer of the skin, and that sounds—pow! Until you remember it's basically what we do already for burns victims. It's what burn therapy is basically about.

The gut is another area which definitely needs to be addressed in this, and it looks again like a pretty ambitious thing until you get to the right literature and you find that ten years ago, in mice, an experiment was done that found the gut wall is capable of rebuilding a perfectly functional intestine with all the villi and crypts and so on just from totally disaggregated cells plated onto a denuded intestinal wall. This is something, which, of course, you wouldn't do by opening up a person. You would do it by some sort of endoscopy technique.

But the proof of principle is ten years old. You know, it's something that you can only really dismiss as science fiction until you've read the right experimental literature.

KONDRACKE: Dr. Sprott, what do you think about those ideas?

SPROTT: I think those are very interesting notions. I think, however, the belief that we are all going to sign up for getting rid of all of our tumors and our telomeres and then check in for a stem cell replacement refill once every ten years—

DE GREY: If your choice is dying of cancer, would you make it?

SPROTT: It doesn't matter whether—I mean, in one sense would I make it were it possible? I suppose I might.

DE GREY: You suppose you might.

SPROTT: Do I think it's possible? No. Do I think we are going to do this society-wide?

KONDRACKE: Why do you think it's not possible?

SPROTT: At this point, while Aubrey said there is proof in principle in a mouse gut, we're a long, long way from a mouse gut to doing all human beings.

DE GREY: Twenty-seven years is a long, long time.

SPROTT: So we finally disagree about how much time twenty-seven years is.

KONDRACKE: OK. Let's see. We can prevent a cancer in a mouse gut. Can we eliminate telomeres from the cells of a—

SPROTT: We could repopulate the mouse gut—

DE GREY: That's right. That's the idea.

KONDRACKE: All right, now what about the differences between mice and humans?

DE GREY: Very important. Very, very important. So I am saying that these things that we need to do in humans in order to eliminate all of the aspects of aging that exist in a normal human lifetime can all be developed in mice within the next ten years.

KONDRACKE: Leaves you only seven years left.

DE GREY: Seventeen years.

KONDRACKE: Seventeen years.

DE GREY: That is not the same as saying that in mice we can make mice, we can do the same proportional life extension. We can take, for example, a two-year-old mouse that would normally live to three years and take it to fifty.

If they can only live to five years, it's not the same as that. It's a lot stronger than that. It's probably going to be relatively easy to turn a mouse that appears old into a mouse that lives for five years rather than three, precisely because mice are simpler than us. Mice are still rather bad at aging, and so fixing them up so that they are better at not aging and can live longer is easier than fixing a human up.

But that means simply that getting mice to live a long time has partly a biological or biotechnological value and partly a publicity value. But of course, the more we do with mice, the more people will acknowledge the plausibility of corresponding progress in the foreseeable future in humans.

KONDRACKE: Now, of these seven categories of things, how many of them have been done in mice?

DE GREY: None of them have been done, but I am saying all of them could be done in ten years.

KONDRACKE: OK. And you've got this prize that you are involved with for the first person to produce a five-year-old mouse, right?

DE GREY: That's not quite how the prize works. It first works incrementally like a world record. So if you produce a mouse that's older than any mouse that we've heard of before, then you get some money. The amount of money you get depends, of course, on how big the prize fund is at the time.

You can go to our Web site, methuselahmouse.org, and put some money on the prize with a credit card. If you want to give us a lot of money, you can talk to us privately.

So it depends on the amount of money—

KONDRACKE: How old is the oldest mouse? What is the longevity of mice now?

DE GREY: So if I could just finish the last question, it also depends on the margin by which you beat the previous record. So that's why it could go on indefinitely.

SPROTT: The answer to your question, Morton, is in excess of five years.

DE GREY: The life expectancy—

SPROTT: Already.

DE GREY: The life expectancy of mice at the moment is about three years depending on the strain. But we are talking about a single mouse here, so we are talking about the maximum life span. The best record that we know of the species that we are interested in, which is the one that most laboratory mice work on, the Methuselah mouse, is just one week short of five years. We gave the inaugural prize in June to Andrzej Bartke who published research in Illinois for this mouse, which was a growth hormone receptor knockout that lived to one week short of five years.

KONDRACKE: Now, as I understand it, though, in order to maintain mouse longevity, what you have to do is you have to calorically deprive them, and you have to genetically alter them so that they are dwarf mice.

Now, as I asked in the last debate that we had, who wants to live his life without hamburgers and in a dwarf status?

DE GREY: You are absolutely right. It wouldn't do at all.

KONDRACKE: In order to achieve long life, OK.

DE GREY: And for this reason we are starting a second prize next year. This prize is simple: it's just that your mouse has to live longer than any other mouse before.

The new prize will be called the reversal prize, and the way in which it will be determined whether you win and how much you get if you do, will take into account the age at which you started treating the mouse. The people who want to take part in this competition will get their mice from some registered place like Jackson Labs, for example. The Jackson Labs will certify the age of the mouse at the time that it was delivered to the investigator, and will also notify us at the prize. And that will be defined to be the age of onset of treatment. So the structure of the prize will be such that the later you start, the more credit you get.

KONDRACKE: OK. But is caloric deprivation not still the way to keep mice alive?

DE GREY: Not really, no. Basically, the thing is with starting later on, not much work is done on trying to extend life span of mice that have had nothing done to them until they are age two or so, precisely because nobody has the faintest idea how to do it.

If you start caloric restriction at age two, you get virtually nothing. If you do any of the things we know how to do at the moment at age two, you get virtually nothing in terms of life extension.

KONDRACKE: OK. Go ahead.

SPROTT: I was just going to comment that I was at a meeting last night listening to David Sinclair talk about his ability to increase mouse life span with late-onset caloric restriction.

DE GREY: How late?

SPROTT: As late as two years, so maybe that's a red herring in the whole thing. One of my concerns about where you go with the caloric restriction—

KONDRACKE: Caloric restriction, by the way, is not on your list.

DE GREY: That's right. I don't really think there's much scope for increasing human life by more than a year or two with caloric restriction.

SPROTT: The only thing we are going to get out of caloric restriction that is of any use to humans is to understand how it works.

DE GREY: Yes!

SPROTT: And then use that for some other sort of therapy.

KONDRACKE: And what do we know about what caloric restriction does?

SPROTT: This was my career.

KONDRACKE: OK, proceed.

SPROTT: What caloric restriction does, in part, is to shift metabolism and produce less of that metabolic damage. It may do a number of other things; it may have significant effects on mitochondrial DNA, for example. It may shift certain kinds of enzymes into a more favorable state.

But clearly, one of the things that concerns me right now is that virtually all of the caloric restriction demonstrations we have are in organisms that live in a very artificial *ad libitum* environment which probably is overfeeding.

Now, that may be the equivalent of the human couch potato myself, obviously, included.

Caloric restriction then isn't going to be of much use directly for us. And whether it's a good model for what changes in humans would be is a very open question, too. I doubt Aubrey disagrees with that.

DE GREY: You are quite right; I don't disagree.

KONDRACKE: Do we have any questions from the audience? Go ahead. There's a microphone up there.

AUDIENCE MEMBER: It seems to me that your arguments are actually not too far apart. I'm kind of surprised about that. You may not agree with me, but it seems like you are in violent agreement except on the amount of time that you would get in maybe ten or twenty years' worth of research.

So the question I have is, or it's a statement at first and then a question, is that when AIDS first became recognized it was a disease that only pariahs in society at that time could have. It seemed to me that was the mental state of the country. And then the people who saw their friends dying started a grass roots effort and got Hollywood involved, and a tremendous amount of money was spent shortly thereafter due to convincing Congress that this was a serious problem, not just among unfortunate communities that few people seemed to care about.

So to make a long question short, why wouldn't the war on aging galvanize the same kind of activity and force cures, because AIDS, which was considered to be inevitably fatal, is now becoming tractable? It's coming into hand as far as treatments.

SPROTT: I'd like to respond to that. I think your questions are reasonable, if one assumes that aging is a disease, and I don't think it is. I think aging is a reflection of a number of basic processes. And I, by the way, did not agree that significant life span extension was possible. I didn't disagree about the amount. I disagreed about the possibility. Aubrey?

DE GREY: Yeah. I think that whether aging is or is not a disease is a red herring. Aging is, first of all, undesirable. And it's a biological phenomenon, so if you want to define disease that way you are done.

But also the main thing is that aging is a process that makes some people more likely to die in a given period of time, starting now, than other people. And that seems like a fine definition of disease to me, you know.

If the only definition of what is a disease and what isn't is terminological, then it's not useful to tell us whether we can fix it or not.

KONDRACKE: Well, what kind of investment would it take, do you think, to achieve your goal?

DE GREY: Well, I normally talk about the ten-year goal. In other words, fixing all these things in mice as a proof of concept, so to speak. And I reckon that Dick's got the money he needs. Twenty million dollasrs a year is a pretty damn good start.

If I had $100 million a year I'd probably think of ways to spend it, but I think with $10-20 million a year, it would probably take only ten years with a good probability.

KONDRACKE: And then how much longer to translate that?

DE GREY: Of course, then we don't know because we don't know how far the science will have progressed for humans in those ten years. It may be that we will be able to leverage off stuff that's gone in parallel, or it may not. We may have made little progress because, until that time, there may continue to be skepticism in society with regard to whether this is ever going to be possible.

So it may be only at that ten-year point that we start putting serious money into things like getting gene therapy to work well in humans.

KONDRACKE: Let me ask the philosophical question: why does anybody want to live to be 5,000 years old? Assuming that you could live to be 130 and be reasonably healthy, presumably you would have to work until you were 100 or maybe you would have to work until you were 80. Do you think that people really want to do that?

DE GREY: Yeah. I think this is another big red herring. We only talk about the desirability of fixing aging because we have the luxury of it being something that most of us feel is a long way away.

People don't want to die when they are healthy. People who get the most of their life might want to retire. They might want to change jobs. They might want to do different things.

KONDRACKE: They want to retire, is what they want to do?

DE GREY: They want to retire because they are getting tired.

KONDRACKE: Um-hmm.

DE GREY: They haven't got the vivacity that they used to have, by and large.

When people have the vitality that they had when they were twenty-five, they are not going to want to sit in front of the television or play golf twenty-four hours a day. They are just not going to.

KONDRACKE: Um-hmm. OK. I'll allow you to start the finish and then Dr. de Grey to finish. We've got about three minutes left.

SPROTT: One of the things I thought was rather interesting, Aubrey, in the very beginning of his presentation, made reference to Jay Olshansky and his point of view. And I think it's rather interesting, Aubrey and I both were among the fifty some scientists who signed a position paper authored largely by Jay, and I thought I might quickly sum up by mentioning some of the key points that we both agreed to when we signed on as co-authors.

KONDRACKE: Quickly.

SPROTT: Yeah, there are about twenty of them, so let me pull out a couple of them that are particularly interesting. "Past and anticipated advances in geriatric medicine will continue to save lives and help to manage degenerative diseases associated with growing older, but these interventions only influence the manifestations of aging, not aging itself."

"Medical interventions for age-related diseases do result in increased life expectancy, but none have been proven to modify the underlying processes of aging."

And finally let me just speak to the last one, which we did both sign. "Although it is likely that advances in molecular genetics will soon lead to effective treatments for inherited and age-related diseases, it is unlikely that scientists will be able to influence aging directly through genetic engineering."

DE GREY: OK. I think that's a good place for us to end because I have an important answer to give to that.

First of all, there is a slight inaccuracy in what Dick said. We did not sign up as co-authors. We signed up as endorsers. And when I was asked to sign up, I took that to mean, "Do you agree with the general thrust of the article?" not "Do you agree with every word or every phrase?" in the way that I would have thought if I were going to be a co-author.

This is an important distinction because certainly the things that Dick has just read out are things with which I do not agree, and which I tried to get changed before publication.

And I wasn't the only one who signed up despite not being able to agree with everything.

What I did agree with was the main thrust of the paper, which has to do with the inability with current technology to do very much about aging, despite the protestations to the contrary from those who make money out of things that don't work.

And I think it's very important for the public to have a good sense of proportion about the efficacy of things we already have, or rather, the lack of efficacy.

So that's why I signed up. But I have taken the trouble to make my views very clear in papers that I have been the author of. For example, I had a paper out two months ago in *Experimental Gerontology* in which I started off by pointing out that it is extremely unfortunate that those who wish to voice legitimate claims that current antiaging medicine, as it's called, doesn't work, tend to spoil their argument by talking about what we might be able to do in the future in the same breath. I think these things are separate. I have very different opinions about both of them, but that didn't stop me from endorsing rather than co-authoring this paper.

KONDRACKE: Thank you both very much, and I am sort of sorry that we are not going to be around to 2030 to find out whether you are right.

DE GREY: Or maybe we will be. Perhaps we will.

KONDRACKE: Thank you very much.

End.

SAGE

crossroads

SAGE Crossroads is an initiative of:

Alliance
FOR AGING RESEARCH

Alliance for Aging Research
2021 K Street NW
Suit 305
Washington DC 20006
www.agingresearch.org

AMERICAN
ASSOCIATION FOR THE
ADVANCEMENT OF
SCIENCE

Advancing science • Serving society

American Academy for the Advancement of Science (AAAS)
1200 New York Ave. NW
Washington DC 20001
www.agingresearch.org

0-595-31631-X